The Burgess Shale

The
Burgess Shale

HARRY B. WHITTINGTON

Published in association with the Geological Survey of Canada
by
Yale University Press
New Haven and London

Designed by James J. Johnson
and set in Caledonia type.
Printed in the United States of America by
Murray Printing Company, Westford, Massachusetts

Library of Congress Cataloging in Publication Data

Whittington, H. B. (Harry Blackmore)
    The Burgess Shale.

    Bibliography: p.
    1. Invertebrates, Fossil.   2. Paleontology—Cambrian.
3. Paleontology—British Columbia.   I. Title.
QE770.W6 1985   560'.1'72309711   85-2297
ISBN 0-300-03348-6 (alk. paper)

The paper in this book meets the guidelines for permanence and durability of the Committee on Production Guidelines for Book Longevity of the Council on Library Resources.

10   9   8   7   6   5   4   3   2   1

To Dorothy

Enthusiastic collector of fossils and
patient helper in all my endeavours

Contents

Figures

Introduction

Fossils, literally things dug up out of the earth, provide the only evidence we have of the history of life. Most fossils are but portions of long-dead organisms—a bone, a shell, a piece of wood—or merely a trace of activity such as a footprint or a burrow. Many shells are incomplete, entire skeletons are rare, and to find traces of soft parts is most exceptional. So the history of life has to be understood from a fragmentary record that is found sporadically in rocks on the continents of the world that were formed during a span of hundreds of millions of years. A discovery of flattened but complete animals in a distant period, when animals lived only in the seas, sheds a uniquely brilliant light on one episode in earth history.

This book is about such a discovery, made early this century in the Rocky Mountains of Canada, by Charles D. Walcott, while investigating rocks that he called the Burgess Shale. A variety of these wonderful fossils were made known by him, particularly through his remarkable photographs. During the next thirty-five years discussion continued on the nature of these ancient animals and their evolutionary significance. Meanwhile the enormous collection Walcott left in the United States National Museum of Natural History lay virtually undisturbed. The quarry he made by digging for them was still there on the mountainside. Occasionally it was visited by geologists, and the ridge in which it was cut was depicted (in reverse) in the engraving on the back of a ten-dollar note of the Bank of Canada, in a series now out of circulation. By the early 1960s the time was ripe for a reexamination of the fossils and the site they came from, using the new tools available.

From time to time I had looked over the collection from the Shale, made by Percy E. Raymond, in the Museum of Comparative Zoology at Harvard University. When the late Armin A. Öpik, a famous investigator of Cambrian rocks and fossils, visited me there, we examined these fossils together. He urged on me the importance of a re-study, so when next I visited my friends on the staff of the Geological Survey of Canada, I broached the idea of such a project. Thus it came about that, under the aegis of the Geological Survey of Canada, I had the privilege of joining in quarrying the Burgess Shale. A large new collection was obtained, and a critical new study has been made of these specimens and of those in Walcott's collection. This book summarises what my colleagues and I have learned, after fifteen or more years, about the fossils from the Burgess Shale. An appendix provides a list of the names of the fossils, with synonyms, and a list of publications on the Shale. The work by authors mentioned in the text will be found in this list, and will provide more detailed information on particular topics. New reconstructions of the animals from the Shale, even models of them, have resulted from our work, and a clearer understanding of why, how, and where they came to be so exquisitely preserved. Hence a far more detailed and lively diorama may be made of the community of organisms in their original environment. The animals are mostly only a few centimetres in length, so that enlarged photographs are necessary to show some of their wonders. They may not be the stuff of a spectacular museum display on the scale of displays of the land-dwelling dinosaurs, but they tell of a past three or four times as old, and are a far rarer find. Their world in the Cambrian period was remote in time and strange compared to ours, and I have sketched some of its relevant features before embarking on the story of Walcott and his discovery, and of the work since then.

I am indebted to many colleagues and various institutions for making work on the Burgess Shale fossils possible. Digby J. McLaren, formerly Director of the Institute of Sedimentary and Petroleum Geology, Geological Survey of Canada, and later Director of the Geological Survey, provided the initiative and support that led to a Survey party collecting at Walcott's original site. McLaren invited me to join the party and to take charge of the palaeobiological work. All members of the party are grateful to James D. Aitken for his inspiring example and experienced, thoughtful, and cheerful leadership of our camp. For study of the fossils, access to Walcott's magnificent original collection was essential. Successive chairmen of the Department of Paleo-biology, US National Museum of Natural History, Smithsonian Institution, Washington, DC, have given such access and provided facilities. Frederick J. Collier, Collection Manager, has been most patient and helpful with numerous loans of specimens. The initial stage of the project was supported by the US National Science Foundation. After I moved from Harvard University to the University of Cambridge in 1966, the palaeobiological work was continued at the Sedgwick Museum, supported by the Natural Environment Research Council and facilities of the Department of Earth Sciences at Cambridge.

I have been encouraged throughout by the enthusiasm and generous help of my colleagues in the palaeobiological studies, Derek E.G. Briggs, David L. Bruton, Christopher P. Hughes, and Simon Conway Morris. They kindly read a first draft of this book and made valuable comments; they also contributed many photographs. Professor J. Keith Rigby has allowed me to use the list of species of sponges he recognises, an original drawing, and photographs, and Professor James Sprinkle furnished two photographs of echinoderms. In the captions to the illustrations acknowledgement is made of this generous help, and of photographs supplied by the Smithsonian Institution and the Geological Survey of Canada. Various drawings and diagrams have been reproduced from scientific journals; where no acknowledgement is given, photographs and diagrams are my own. Aitken and William H. Fritz of the Geological Survey of Canada, and Desmond H. Collins of the Royal Ontario Museum, Toronto, have kindly answered queries. The Superintendent of Yoho National Park, Parks Canada, sent me copies of literature relevant to the Burgess Shale site. Edward Tripp, Editor-in-Chief, Yale University Press, encouraged me to write this book, and has helped me in many ways. Miss Adele Prouse produced the final drawings, and Mr David Bursill made enlargements from my and my colleagues' negatives. The patient and accurate typing of successive drafts of the text by Mrs Sandra Last is also gratefully acknowledged.

1 *The Cambrian World*

The Burgess Shale was formed some 530 million years ago, during the Cambrian period in Earth history (table 1.1). In this period the world was very different from today. The geography (fig. 1.1) has many uncertainties, but four major continents are recognised. A continent in this sense includes the area that was land, the adjacent area of the continental shelf covered by shallow sea, and the slope at the margin of the shelf. The continent named Laurentia was comprised of present-day North America and Greenland, with the addition of portions of northern Britain, western Spitsbergen, and western Newfoundland. The Cambrian and early Ordovician rocks and fossils of these three last-named areas are so like those of North America that it is thought that they were deposited in one contiguous region. For similar reasons eastern Newfoundland, portions of New Brunswick, Massachusetts, and Rhode Island are thought to have been part of West Eurasia, a continent that included Europe north of the Alpine Mountains, and that extended to the line of the present Ural Mountains. The large continent of Gondwanaland comprised present-day Africa, South America, Australia, Antarctica, Arabia, peninsular India, and Madagascar. East Eurasia may not have been a single continent but a number of smaller ones that would have included the present areas of China, Kolyma, Siberia, and Kazakhstan.

Because there is considerable uncertainty about the relative positions of the continents, the gaps between them may have been greater or smaller than shown, so that the width of the seas that sepa-

	Era	Period	Millions of years before the present
Phanerozoic	Cenozoic	Quaternary	
		Tertiary	66
	Mesozoic	Cretaceous Jurassic Triassic	245
	Palaeozoic	Permian Carboniferous Devonian Silurian Ordovician	
			500
		Cambrian	570
		Precambrian	

Table 1.1. The names of the major divisions of time used in Earth history and estimates of their duration. Phanerozoic (from *phaneros*, a Greek word meaning visible) is a convenient term for the time during which evidence of life was abundantly preserved. The names of the Eras into which the periods are grouped reflect the broad pattern of life revealed—Palaeozoic, or ancient, life; Mesozoic, or middle, life; and Cenozoic, the newer forms of life that merge into the present. The name Precambrian refers to the various rocks formed in the immense length of time before the beginning of the Phanerozoic. The oldest rocks known so far were formed about 3,800 million years ago, and the earth's origin is dated at about 4,600 million years ago. These dates and the estimates of ages of particular boundaries in the table are derived from the study of radioactive isotopes in rocks. Each period is the length of time during which a system of rocks was deposited; the name of the period is the same as that of the system. The names of the systems were adopted as the science of geology developed, and are derived from the names of areas in which the rocks were first studied (for example, Cambria is the ancient name for Wales) or the conspicuous rock in a particular system (for example, *creta* is Latin for chalk).

rated them is uncertain. There is no evidence of a land mass at either pole, hence there were no ice caps, and Cambrian climates may have been less extreme than present-day ones.

The sedimentary rocks that formed on the Cambrian continental shelves provide a record of the events of the period. Where such deposits are unknown today (in the stippled areas of fig. 1.1), they have either been eroded away or were never formed. These areas may have been land. There is no evidence of plants or animals on the land areas, or in fresh or brackish water at their margins, so we have to imagine

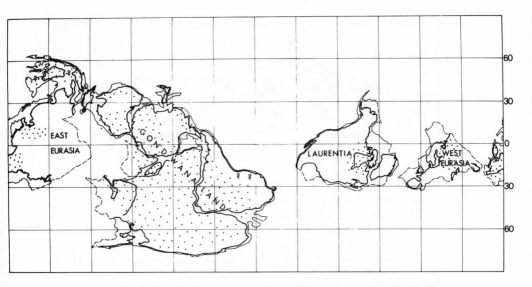

Fig. 1.1. A possible geography of the world in Middle Cambrian times. The positions of the continental areas are based on measurements of the direction of magnetism preserved in samples of rocks of Cambrian age. There is great uncertainty about the relative positions of the four named continents. Present-day continents, or portions of them, are shown in heavier outline, the outlines of possible Cambrian continents in thinner line. Stipple shows areas of these continents where Cambrian rocks are unknown today. During the Middle Cambrian the unstippled parts of these continents were probably covered by sea, in which rocks were deposited. From Whittington, 1981, *Proceedings of the Linnean Society, New South Wales*, vol. 105, p. 83, fig. 2.

Cambrian lands as barren, uninhabited, subject to erosion, and as supplying sediments to the surrounding seas. Such was the case with the land area of Laurentia, centred on the present Hudson Bay. The Cambrian sandstones, siltstones, and mudstones deposited around this land all show features characteristic of formation in shallow water. Farther away from the old land, Cambrian rocks are dominantly limestones, formed in clear, shallow waters beyond the limits to which most land-derived sediments were carried. The position of Laurentia relative to the equator means that these waters were warm. Farther away, towards the borders of Laurentia, Cambrian rocks in what are now portions of the Rocky Mountains and the Appalachians are dark shales and thin, dark-coloured muddy limestones. The muddy sediments appear to have been brought from some source other than Laurentia. The cross-section (fig. 1.2) shows the different kinds of deposits

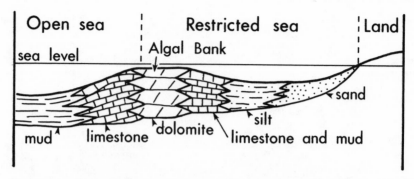

Fig. 1.2. A diagrammatic cross-section of the different kinds of deposits that accumulated in the shallow seas (0–300 m deep) off the western coast of Laurentia (part of which is now North America—see fig. 1.1). The vertical scale is much exaggerated. In the sand, salt, mud, and limestone deposits of the seas on the landward side of the algal banks are found species of trilobites that were limited in geographical distribution, because the algal banks restricted circulation to the outside. In the warm climate of low latitudes, this restriction may have led to an increase in salinity in the shelf seas. In the more open seas outside the algal banks, in the muddy and limey deposits, different species of trilobites are found, including agnostoids. These species are more widely distributed geographically because of migrations from and to other areas. The central portion of the algal bank was originally of calcium carbonate deposited by the algae, but was altered shortly afterwards to dolomite (calcium-magnesium carbonate). The Burgess Shale was deposited at the margin of the open sea. This facies diagram illustrates the different environments of deposition in which various groups of species lived contemporaneously (the term 'facies' embraces the type of rock formed and the fossils it contains). Changes of sea level with time would cause sideways shifts of environments; for example, a rise in sea level might cause the algal banks to be covered by open sea muds. Adapted from a diagram by R.A. Robison, 1976, *Brigham Young University Geology Studies*, vol. 23, pt. 2, p. 94, fig. 1.

being formed contemporaneously in the various environments progressing outwards from the land. The bank was built up by algae and other organisms in the shallow, well-oxygenated waters of the continental shelf that were free from land-derived sediments; muds accumulated in the deeper open sea waters outside the bank. The shallow bank limited circulation between the open and restricted seas.

Sea level during the Cambrian fluctuated in relation to the Laurentian continent. From fig. 1.2 it may be seen that a lowering of sea level would have reduced, or even terminated, marine deposition landward of the bank, until the level rose again. A relative rise in sea

level would have renewed such deposition and may have resulted in
muds from the open sea being brought in over the bank. Such fluctua-
tions caused interfingering of the restricted and open-sea deposits
during the Cambrian, particularly adjacent to the algal bank.

The sedimentary rocks formed on the shelf of Laurentia show that
the physical and chemical conditions affecting their origin were not
radically different from those of today. The biological factors were not
similar, however, for Cambrian fossils show how different the organisms were from those on present continental shelves. Had a modern
shell-collector been able to stroll along a Cambrian beach, the pick-
ings would have been very different: he would have found no heaps
of varied bivalve and snail shells, no fragments of bryozoan colonies or
corals, and no signs of bony fish or sharks. Nevertheless, there would
have been heaps of shells on the beaches, for shells are found commonly in Cambrian rocks on all continents. These shells (composed of
calcium carbonate, phosphatic material or the tough organic material
called chitin) were not decomposed but instead were entombed in
either a complete or a fragmentary state. They may have been buried
in the sediment accumulating on the shallow sea floor where they
lived, or they may have been carried with these sands, silts, or muds
to the site of final deposition. In these shallow margins of the continents life was abundant and the waters well oxygenated, as is the case
today. Chemical decomposition, in the presence of bacteria, and the
action of scavenging animals, would have removed all trace of the soft
parts before or soon after burial. Thus, only in some exceptional circumstances, could soft parts be preserved.

The fossils most frequently found in large numbers in Cambrian
rocks are the mineralized shells of trilobites (figs 1.3–6). Also common
in Cambrian rocks are cap-shaped or flattened shells. Many of these
are brachiopods, classed as inarticulate (see figs 4.11–13) because the
two parts, or valves, of the shell were not connected together, but
were held in place by muscles. Most inarticulate brachiopod shells
were formed of organic and phosphatic material. When such shells are
enclosed in limestone, it is easy to free them by dissolving the matrix
in dilute acetic acid, which does not affect the shells themselves.
Many beautifully preserved shells have been extracted in this way.
Articulate brachiopods (see fig. 4.14), which had a pair of teeth in one
valve that fitted into sockets in the other, had calcareous shells. They
are present in early Cambrian rocks, but not as abundantly as in post-
Cambrian rocks. Brachiopods occur today only in the sea; they appear
to have been marine animals throughout their history. The small cap-

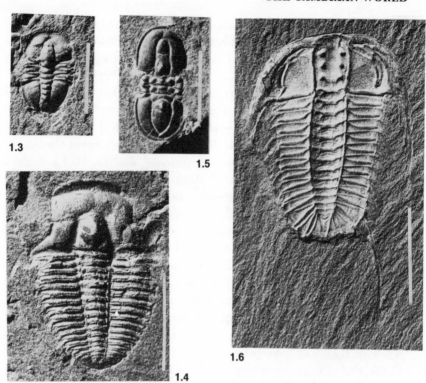

1.3

1.5

1.6

1.4

Fig. 1.3. *Pagetia bootes* from the Burgess Shale. Small trilobites of this type occur in Lower and Middle Cambrian rocks and may have eyes and two or three segments, as in this example. Scale bar 0.5 cm.

Fig. 1.4. *Elrathia permulta* from the Burgess Shale. Trilobites like this, which have many segments and a small tail, are common in the Middle and Upper Cambrian.

Fig. 1.5. *Ptychagnostus praecurrens* from the Burgess Shale, an agnostoid trilobite, typical in having a similar-sized head and tail separated by two segments. Agnostoid species are widely distributed in Middle and Upper Cambrian rocks. Scale bar 0.5 cm.

Fig. 1.6. *Oryctocephalus reynoldsi* from the Burgess Shale, a representative of a world-wide family of Lower and Middle Cambrian trilobites.

shaped (see fig. 4.17) and coiled shells that also occur are early kinds of molluscs, but molluscan bivalves and snails like those common to-day are hardly known from Cambrian rocks.

Careful searching in Cambrian rocks may reveal the needle-like spicules of sponges, in some forms arranged in a rectangular pattern

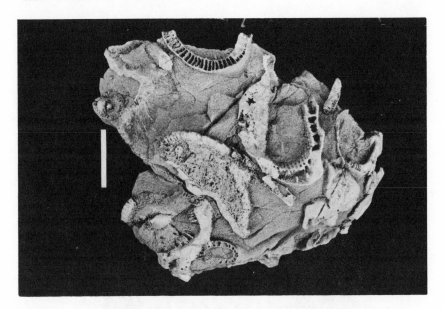

Fig. 1.7. Archaeocyathids in a block of Ajax Limestone from South Australia. The archaeocyathid skeleton, originally of calcium carbonate, has been replaced by silica and is weathered out in relief so that it shows the double walls separated by radially-directed partitions. Specimen in Museum of Comparative Zoology, Harvard University.

(see fig. 4.9). The network of spicules, which were either siliceous or calcareous, supported the walls of the sponge. In many of these early sponges the spicules were joined together only weakly, so that they became separated after death and decay of the soft parts. Thus, though the occurrence of thin rock layers packed with spicules shows that these animals were abundant and varied in the Cambrian, localities yielding entire specimens are rare. Possibly related to sponges were the Lower Cambrian archaeocyathids, which occur in masses in some limestones (fig. 1.7). Their conical, calcareous skeletons were built in most species of an inner and outer wall, connected by vertical, radial partitions, but empty in the centre. The walls and partitions were perforated, the skeleton reduced to a network of rods in some examples. Since no soft parts are preserved, and the skeleton is not exactly like that of a sponge, the affinities of this group have been debated. Their occurrence in lens-like masses, some upright, the spaces between filled with broken parts of skeletons, suggests that they lived gregariously, attached to the sea floor or to dead skeletons, and formed small patch reefs and banks in shallow, clear waters. In

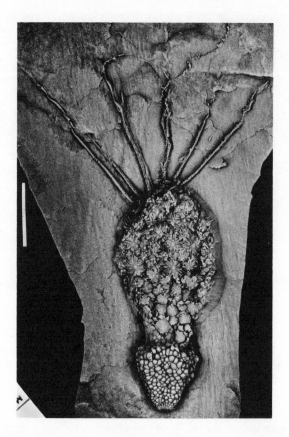

Fig. 1.8. *Gogia guntheri*, a Middle Cambrian echinoderm from Utah. The conical, plated structure at the base was a holdfast to keep the animal in place on the sea bottom. The oval body was covered by plates that had pores at the margin through which water was circulated for respiration. The arms were flexible, spiral distally, and bore hair-like processes that created currents to carry food particles along the arms to the central mouth. Photograph by J. Sprinkle.

similar conditions, but not in all cases in association with archaeocyathids, blue-green algae (or cyanobacteria) built up laminated mounds of calcium carbonate. The successive layers of lime were built up as a result of the biological activity of the algae. Some of these algal mounds, known as stromatolites, formed extensive layers one or two metres thick, and in particular circumstances are thought to have formed banks bound together sufficiently firmly to have been wave-resistant and thus reef-like.

Sea-lilies (crinoids), starfish, and sea urchins of today are exclusively marine in habit, and have internal calcareous skeletons, the characteristic microstructure of which is unique to echinoderms. Cambrian rocks have yielded calcareous plates showing this unmistakable echinoderm microstructure, and hence must have been formed in the sea. Entire skeletons (fig. 1.8) are rare finds. Such exceptional specimens, including those from the Burgess Shale described here, reveal that half the twenty major divisions (or classes) of echinoderms were present in the Cambrian. Eight of these classes, however, became extinct during the Palaeozoic Era, only the crinoids (see fig. 4.64) and sea-cucumbers (see figs 4.62, 4.63) surviving from the Cambrian to form major living groups.

This brief survey shows that shells from Cambrian rocks all belong to only five major groups (see fig. 5.2) of marine animals. The variety of kinds within each group is considerable, especially of trilobites and echinoderms. The assemblages of shells found in particular places, and in Cambrian rocks at different levels, are unlike those we find today. They are mostly from trilobites and inarticulate brachiopods, but include also tubular, conical, and cap-shaped shells. The bivalves and gastropods found in abundance in shallow waters today were virtually absent, as were bryozoans and corals, and there were no bony fish or sharks. In warm, shallow waters there may have been small mounds, or patches, built up from the sea bottom by archaeocyathids (Lower Cambrian) or by mats of algae and bacteria, but no great barrier reefs or atolls like those made by stony corals in the present tropics.

The great importance of the Burgess Shale fossils lies in their revealing the incompleteness of the picture of Cambrian marine animal communities given by shells alone. Not only are twice as many major groups of animals represented in the Burgess Shale community (see fig. 5.2), but soft-bodied animals far outnumbered those with hard parts, and were twice as varied in kinds. This proportion of species of soft-bodied animals to those having hard parts is matched in marine communities today, hence it is not unusual. Thus there is good reason to assume that the Burgess Shale fauna more nearly represents what a Cambrian community was like than does an assemblage of fossils from a deposit containing hard parts only. In the Shale, unlike the latter faunas, trilobites were not dominant, but other kinds of arthropods that lacked a mineralized shell were. Additional soft-bodied animals included not only a unique variety of worms, but also the earliest known fossil examples of important later groups, such as the crinoids and sea-cucumbers, and chordates. Other animals in the Shale defy classification within accepted major groups, and present unique problems in evolutionary relationships.

In recent years, intensive collecting of fossils at sites up to 1300 km south of the Burgess Shale locality has yielded specimens of a variety of similar or identical species of the soft-bodied animals. This marine community was thus widely distributed in the Laurentian seas, and was not the outcome of evolution in some unique local environment. Rather, soft-bodied animals have been found because of the exceptional conditions of preservation, nowhere more so than at the Burgess Shale site.

More clearly than at any other locality where Cambrian fossils are found, those from the Burgess Shale reveal the evolutionary stage that marine life had reached, in a strange world of the remote past. How much they add to the knowledge of this life may be appreciated when the fossils described briefly in this chapter, which are those found under usual conditions of preservation, are compared with those described in chapter 4. How this window on the past was found, how it came to be there, and the nature and significance of what it shows us, are the subjects of other chapters.

2 Walcott's Discovery and the Subsequent Work

After more than forty years of searching for Cambrian fossils all over North America, Charles D. Walcott made his greatest discovery in August 1909. He and his party, which included his wife and two of his sons, were searching for fossils in the neighbourhood of Burgess Pass (fig. 2.1) before returning home from a long season's exploration in the Canadian Rocky Mountains. Charles Schuchert, in his obituary of Walcott, recounts the story that Mrs Walcott's horse stumbled as the party were making their way southwards along the trail (fig. 2.1) to the railroad. Walcott dismounted and split a slab of rock lying loose on the trail. It revealed a magnificently preserved fossil, a shining silvery film on the dark rock. Walcott's diaries, however, show that he and his party had camped below the ridge between Mount Field and Wapta Mountain (fig. 2.2) for two or three days, and had been searching for fossils on this ridge. On 31 August, he records that they made the great discovery by splitting a loose block that he and Mrs Walcott found. It contained specimens of what they informally called the "lace crab", later named *Marrella splendens* (see fig. 3.3). Whatever the exact circumstances, the discovery was the reward of long and assiduous searching for fossils, and one of the most outstanding ever made.

Walcott's experience enabled him to recognize the significance of his find, namely, that soft parts, such as limbs, were preserved. Before him lay the enticing prospect that, if he could locate the exact layer from which this block had come, a major discovery was probable. So the next season he came back, and he and his two sons searched the

11

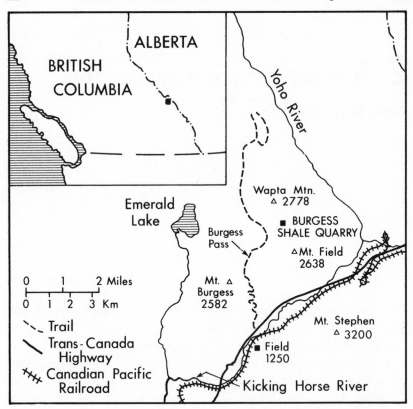

Fig. 2.1. Map of part of Yoho National Park, British Columbia, Canada, showing the location of the Burgess Shale quarry on the west slope of the ridge between Wapta Mountain and Mount Field. Heights are given in metres.

steep slope above the trail until they found the fossil-bearing layers. These were within about two metres thickness of the 150 m of the Shale. It is fortunate that at this spot the layers are almost horizontal, so that they could be quarried by digging into the steep slope. Thirty days were spent quarrying in 1910, and longer in the following year. By then a quarry 20 m long, extending 3 m back into the ridge, had been developed, and 120 m³ of shale split for fossils. The quarrying, using picks, chisels, long iron bars, and small charges of explosive, was continued in 1912 (fig. 2.3) and 1913, and in 1917 a final fifty days were spent there by Walcott, then in his sixty-seventh year. So was amassed the unrivalled collection of over sixty thousand of these unique fossils now in the US National Museum of Natural History, Washington, DC.

Fig. 2.2. The ridge between Mount Field and Wapta Mountain (left) showing the sites of the Walcott quarry (W) and the excavation made by Raymond (R). Below the trail (T), shown also in fig. 2.1, is the camp (C) of the
Geological Survey of Canada party in 1966 and 1967. The dashed line on the
ridge north of the Walcott quarry shows the abrupt change between the
dark Burgess Shale and the light-coloured limestones and dolomites of the
Cathedral Formation. 9 August 1967.

A brief biography of Walcott will explain how his unique collection of Canadian fossils came to the United States. Walcott was born
in 1850 in Oneida County, in northern New York State. His formal
education, in the public schools of Utica and at the Utica Academy,
lasted only ten years. His father died when he was young, so he left
the Academy in 1868 and spent the next two years as a clerk in a
hardware store, a practical business training that was to stand him in
good stead. While at school he had become interested in natural history, collecting fossils, birds' eggs, and insects, and had been greatly
influenced by Colonel E. Jewett, a geologist and palaeontologist who
had lent him books and taught him how to collect. When business
took him to Indianapolis in 1871, he met E. T. Cox, who was making
a geological study of Indiana coalfields. This meeting helped him to
decide to leave business and devote himself to geology. Returning to
Trenton Falls, New York State, he lived and worked part-time on the
farm of William P. Rust. In his free time during the next five years he
made an important collection of fossils, mainly from the Ordovician

Fig. 2.3. Charles D. Walcott (probably in 1912) standing on a fossiliferous layer in the Phyllopod bed at the south end of the quarry. He is holding one of the long iron bars used in levering out slabs of rock. Photograph courtesy of Smithsonian Institution Archives, Charles D. Walcott Collection, 1857–1940.

Trenton limestones exposed in the deep river gorges nearby. He sold this collection in 1873 to Louis Agassiz, the great Swiss naturalist who had founded the Museum of Comparative Zoology at Harvard University. Walcott's hopes of studying with Agassiz were dashed when Agassiz died, but he continued his work, particularly on trilobites, and published four papers on them in 1875. The next year he received his first professional appointment, as assistant to James Hall, the famous palaeontologist and State Geologist of New York. From Albany he moved to Washington, DC, to join the newly-formed US Geological Survey in 1879.

Walcott's career in the Survey was remarkable, for he was a vigorous, decisive, and commanding personality, with great administrative ability. He was in charge of the palaeontological work by 1883, and in 1894 was appointed Director of the Geological Survey by President Cleveland, in succession to Major J. W. Powell. As a young man Walcott had found fossils in rocks older than Ordovician, and this discovery led him to take up the study of those older fossiliferous rocks that the English geologist Adam Sedgwick had called Cambrian. In his work for the Geological Survey he studied these rocks, describing the succession of strata, and particularly the trilobites and brachiopods, from Quebec and Newfoundland to Alabama in the eastern part of the American continent, and in Texas, California, Nevada, and other western areas. He became a world authority on the Cambrian System of rocks, visited Europe, and studied Cambrian collections from India and China. While Walcott was Director of the US Geological Survey, his administrative talents enabled him to expand the staff and extend the work into water resources, land reclamation, and protection of forests.

Only two years after becoming Director, he had also become Assistant Secretary of the Smithsonian Institution, in charge of the US National Museum, the repository for collections of fossils made by the US Geological Survey. Walcott was thus able to continue his scientific work on the Cambrian. His drive, energy, and authority resulted in his playing an important part in founding another scientific centre, the Carnegie Institution. His administrative talents profoundly influenced not only the Survey, but also the US National Museum and the formative stages of the Carnegie Institution. It was early in 1907 that he was persuaded to accept the Secretaryship of the Smithsonian Institution, and he left the US Geological Survey later that year. This new post enabled him to continue his scientific work, and also turn his attention in the summer months to the Canadian Rocky Mountains. His explorations, by pack horse, led to the first understanding of the enormously thick Palaeozoic rock section, Cambrian to Devonian in age, in these mountains. It included, as we have seen, his major discovery of the Burgess Shale. Walcott was not the first to find fossils in this area. While surveying the section for the railway in 1886, Otto Klotz, Director of the Dominion Observatory, had found trilobites and other fossils near Field, low on the slopes of Mount Stephen (see fig. 2.1). No doubt the published descriptions of these fossils were what drew Walcott's attention to this particular place, where he made a large collection, and to the adjacent region. Klotz, in the course of his survey, named mountains in the region, including one for A. M. Burgess, then Deputy Minister of the Interior for Canada.

Walcott was a decisive and hard-working scientist, who brought back his collections each summer and studied them in the winter, producing a steady stream of accounts of the fossils and their significance. In 1911 his second wife was killed in a railway accident, and his oldest son died while a student at Yale University; further personal tragedy was the loss of another son, shot down in France in 1917. Walcott was heavily engaged in administrative work and in aeronautics during the First World War. Despite the burden of his duties and the personal tragedy, Walcott was able to describe his finds from the Burgess Shale in a series of scientific articles published between 1911 and 1924 (see Appendix). He was fifty-nine years of age when he made his discovery, and it was a truly remarkable feat to both make his collection and publish an account of it in the next fifteen years. Walcott wrote a large number of reports and scientific articles on other topics during those same years, continuing until his death in 1927. A posthumous article on some Burgess Shale fossils, published in 1931, was prepared by his colleague, Charles E. Resser, from notes that Walcott left.

The photographs with which Walcott illustrated his publications were made mainly by reflecting light from the shiny surface of the thin layer in which the fossils are preserved. They revealed in amazing detail hitherto unknown worms, algae, sponges and arthropods, and created a sensation among both palaeontologists and zoologists. Walcott had friends in many countries. In museums in North America, Europe, and Australia, I have seen small collections of these precious fossils that he distributed—perhaps to help convince people of their reality! He also left, in his Annual Reports to the Smithsonian Institution, a fascinating record of his field work and how it was done. An expert photographer, he captured splendid images of the Rocky Mountains and of work in camp. His third wife, Mary Vaux Walcott, whom he married in 1914, accompanied him on his explorations in the Canadian Rockies. She painted the wonderful and varied wild flowers of the region, some common and others so rare that she saw them only two or three times in the course of long rides above the tree line. Her book, published by the Institution, had 400 plates of her paintings of flowers, mainly from the Rockies, observed at heights of up to 2300 m.

Walcott's entire collection from the Burgess Shale was made readily accessible in the US National Museum when the new wing was completed in 1963, and the palaeontological collections were reorganized in new quarters. Prior to that I had photographed many of the specimens Walcott had illustrated, but these were only a small proportion of the total. At Harvard University there was also a considerable collection of specimens obtained by Raymond. In 1930 Raymond had taken three of his students to the Canadian Rockies and they had spent about three weeks at Walcott's site, shovelling the snow out of the quarry and drilling and blasting to get out some of the rock. Because of the way in which they were obtained, the specimens in both Walcott's and Raymond's collections are vaguely labelled as having come from the quarry, or from a level about 21 m above it, but exactly where, to the nearest few centimetres, is unknown. Walcott did record a measured section in his quarry, indicating that the exceptionally preserved fossils came from a few thin layers, and that the species found in a particular layer were not all the same, but individual specimens are not labelled accordingly. Further, how the rocks in, above, and below the quarry—the dark, fine-grained Burgess Shale of Walcott—are related to those adjacent and in the general area, was unknown.

Since the late 1940s, however, the Geological Survey of Canada has been pursuing a vast mapping programme, which in the mid-1960s embraced the southern Rocky Mountains of Alberta and British Columbia. Earlier I had discussed with friends on the Survey the desirability of a re-investigation of Walcott's site, and the possibility that a collection of the famous fossils might be obtained for Canada. In 1966, this discussion bore fruit, and Aitken of the Geological Survey was appointed leader of a party which included his colleague Fritz, an expert in Cambrian trilobites and stratigraphy, and me, to direct the work on the other fossils. Aitken and Fritz, aided by MacDonell, Green, Lambert, Stesky, and Johnson, established the camp (see fig. 2.4), made a trail up to the Walcott quarry, and began to clear the floor of debris. Judith Fritz and son Peter, my wife and I, and James A. Doyle (then a student at Harvard University) joined the party, and we all spent about six weeks in camp. In the following year, a similar party spent a further six weeks continuing work at the north end of Walcott's quarry with considerable success. That year we were joined by Bruton, of the University of Oslo, as well as by Mrs Anne Aitken.

I had read many of Walcott's reports on his work in the Canadian Rockies, so my first sight of his quarry was a great thrill (see fig. 2.2).

Fig. 2.4. The 1966 Geological Survey of Canada party outside the mess tent. From left to right, front row: Norman W. MacDonell, assistant; James D. Aitken, leader; Peter Fritz; Dorothy A. Whittington; James A. Doyle (now a distinguished palaeobotanist); Henry Lambert, assistant. Back row: Robert M. Stesky, assistant to Aitken; Terry Green, assistant; Judith Fritz; H.B. Whittington; Clifford Johnson, assistant to Fritz; William H. Fritz; Riba Nelson, camp cook. Photograph by Mike Kelly, Communications Branch, Department of Energy, Mines, and Resources, Canada.

It came after climbing up the trail through the woods from the town of Field to Burgess Pass (see fig. 2.1). The tents of the camp, below Wapta mountain, looked most welcoming. The trail we were following was the one along which Walcott was riding when he found his first block of shale, and the photographs he had published enabled me to recognise his quarry high on the ridge. On that first expedition, everyone took a share in camp chores and the quarrying operations. Splitting of slabs for fossils was greatly helped by the enthusiasm of the two wives in the party. Most days we all climbed up to the Walcott or Raymond quarry, only our experienced cook, "Ma" Nelson, remaining in camp to produce the excellent meals that kept us going.

The debris that had accumulated on the flat rock floor of Walcott's quarry, part of which had been left by him, and part of which had slid down from the slope above, was a formidable barrier to getting at the 6-m wall of the quarry, especially the lowest 2 m, which yielded the

Fig. 2.5. Members of the Geological Survey of Canada party splitting off a layer of the Phyllopod bed, and shovelling debris from the floor of the quarry. Aitken is on the left, and the circle on the rock, beside his left arm, surrounds the arbitrary mark from which all levels in the strata were measured. 7 August 1966.

fossils we were after. Walcott called this lower 2 m the Phyllopod (literally, leaf-foot) bed because it yielded so many fine specimens of animals having leaf-shaped limbs; four thin layers within this interval were particularly notable for the many excellent specimens they contained. The quarry penetrates deepest into the ridge at the southern end, which is probably where Walcott worked in later years (judging from photographs he left such as fig. 2.3), and the pile of debris at that end is too great to move by hand. Hence we tackled the northern end of the quarry, first clearing the debris from the rock floor. We then cleared an area at the top of the back wall, from which we worked downwards, splitting with chisels and hammers the thinnest possible layers. From an arbitrary mark on this wall (see fig. 2.5) our measurements recording the thickness and position of each layer were made. During sixty days of quarrying in 1966 and 1967, we extended the quarry some 12 m northwards, and in our search for fossils, split about 700 m³ of rock. Each layer was split carefully and each fragment which contained a fossil was marked as to the exact level from which it came, and was wrapped and packed in bags marked with the date of collection, and level from which the contents came (fig. 2.6). Every day we

Fig. 2.6. The north end of the Walcott quarry, partially filled with snow (foreground), showing a layer being split in search of fossils, and in the background labelling and packing of specimens, and a coffee break in progress. 13 July 1966.

carried packs heavy with these bags down to camp. The procedure was similar at the excavation we made about 22 m above Walcott's quarry, a level from which Walcott and especially Raymond obtained good material of a few species. We broke up 17 m³ of rock from this small quarry, which we called the Raymond quarry after its discoverer. Our efforts were concentrated on these two levels in the ridge, because no comparable specimens of soft-bodied fossils were found at other levels, though we explored only briefly up and down the slopes.

Our methods of obtaining fossils were thus not so very different from Walcott's fifty years earlier. Admittedly, he did not have the advantage of the felt-tip pen, to record on each piece of rock when it was collected and the level from which it came. We used blasting, but only in a limited way, to open up vertical cracks so that we could remove the rock carefully layer by layer. Walcott and Raymond probably used heavier charges, so that the layers were disrupted and blown out a short distance, hence exact levels could not be recorded. Pack trains of horses were the traditional means of moving gear and supplies for geological work in the Rockies, but the helicopter has changed this. Comfortable camps can now be set up in the valleys and parties flown out daily—no longer does the geologist face a long climb each day

Fig. 2.7. A packer dismounting as he delivers supplies in the traditional manner. In the foreground is lumber flown in to make boxes; in the rear, the mess tent and adjacent snow bank providing water and refrigeration. 18 July 1966.

before starting work. Our camp was brought in and removed by helicopter. Wooden boxes (made in camp from lumber flown in) containing the collections were also removed this way. We obtained weekly supplies by pack-horse (fig. 2.7). A patch of snow gave us a water supply and refrigeration. The variety and succession of wild flowers were a constant joy, mountain goats were seen occasionally, and we were visited once or twice by humming-birds, perhaps attracted by our brightly-coloured hard hats.

We had to be sure that the packed lunch we took daily was safe from the numerous rock squirrels. Walcott's party, when blasting took place, would retire to a safe distance, and as Mrs Walcott wrote, often used to feed the squirrels. After a gap of several years, they returned to the quarry and resumed operations, and she records how, the first time a blast was fired, the squirrels appeared at once, expecting at this signal to be fed! One evening we searched for the old Walcott camp site. It was among the trees below Burgess Pass, where a small stream from melting snow provided water. The site was dark and rather dismal, if sheltered, compared to our site above the tree line. It could be located readily, by a rusting stove and a meat hook in the trees, a clearing in which horses had been tethered (Walcott used to ride up to the quarry each day and a small trail still exists), and squares where tents had been erected. At this elevation the vegetation is very slow to recover from any disturbance, so these squares were clear and undisturbed, with boughs cut for beds arranged on them. Outside one tent-square was a large pile of chips of shale, for Walcott used to bring slabs down by horse and split them at his tent.

The Burgess Shale outcrop is in Yoho National Park (see fig. 2.1), and camping is not allowed outside designated sites, except by special permission, such as that obtained by the Geological Survey of Canada party. Similar permission was given in 1975 for a party, led by Collins, from the Royal Ontario Museum, Toronto. This party was allowed to collect from the debris in the Walcott and Raymond quarries, and the slopes above and below; it also collected from the piles of debris in Walcott's camp. A large and valuable amount of material was obtained that included some rarities, and even the counterpart of a specimen that had been brought back by Raymond. This collection has provided specimens for display not only in the Royal Ontario Museum, but also by Parks Canada and many other museums in Canada. The illustrated account of the party's work by Collins includes a photograph of the site of Walcott's camp. In July 1981 the Burgess Shale outcrop became a World Heritage Site (one of five in Canada), and a plaque recording this decision stands in the valley below Mount Field. An outdoor exhibit to accompany the plaque is now in preparation, and visits to the site, and to the fossil bed on Mount Stephen, may be made only with a Park Warden. Collecting fossils is not permitted, and the Walcott quarry is under special surveillance. In the Smithsonian Institution's National Museum of Natural History in Washington, DC, a new exhibit has been opened that shows more than eighty of Walcott's fossils, and a fine new diorama portraying the site as it is thought to have been in Cambrian time (see fig. 5.1).

The clearer understanding of where and under what conditions the Burgess Shale was formed, suggested that other sites nearby might yield similarly-preserved fossils. This possibility was explored in 1981 and 1982 by parties again led by Collins, with Briggs and Conway Morris. Their recent publication recounts how they found fossils like the exceptionally-preserved ones from Walcott's quarry at over a dozen new localities in the adjacent area.

3 *Formation of the Burgess Shale and Preservation of the Fossils*

The objectives of the Geological Survey of Canada party included gathering information that would help in understanding the setting in which the Burgess Shale was formed, which in turn might lead towards explaining why these extraordinary fossils were preserved. Walcott named the Shale from the exposed section on the ridge, a section that extends from below his quarry up towards the top of the ridge and is some 150 m thick. In figure 2.2 this section is dark in colour, but a short distance to the left (that is, northwards), and at the top of the ridge, the rocks are light-coloured limestones and dolomites (both these kinds of rock are carbonates, limestone a calcium carbonate, dolomite a calcium and magnesium carbonate, formed in most cases by alteration of original limestones). The change from shale to carbonate rock is abrupt, along a line (dashed in fig. 2.2) directed steeply up the ridge. At the south end of the ridge, overlooking the Kicking Horse River valley, is Mount Field (fig. 3.1), in which is exposed a long section in Cambrian rocks (Lower Cambrian strata underlie the Middle Cambrian shown in the photograph). Again, an abrupt, almost vertical boundary may be traced, separating dark shale of the Stephen Formation (which includes the Burgess Shale) to the southwest from carbonate rocks of the Cathedral Formation to the northeast (on the right of the photograph). It has long been realised that Wapta Mountain, Mount Field, and the ridge joining them, which runs in a northwest to southeast direction, lie approximately on a line of major facies change in the southern Rocky Mountains of Canada. To the northeast

Fig. 3.1. The southeast face of Mount Field, photographed from the side of Mount Stephen, across the Kicking Horse River valley. The exposed Middle Cambrian rocks are shown in fig. 3.2. The trace of the boundary between the Cathedral (C) and Stephen (S) formations, marked by a dashed line, includes the almost vertical edge of the escarpment. Above is the boundary between the Stephen and overlying Eldon (E) formations. The position of some of the massive reef core rocks (R) is indicated. 11 August 1972.

of this line the Cambrian rocks are largely carbonates and sandstones, in nearly horizontal sheets, resistant to weathering and magnificently exposed in steep mountain cliffs. To the southwest the rocks of the same age are shales and thin limestones, less resistant to erosion and forming lower mountains that are covered with trees. This major facies change is that described in chapter 1 (see also fig. 1.2), between the rocks formed in the shallower seas covering carbonate banks, and those formed in the deeper waters off-shore, where dark muds and

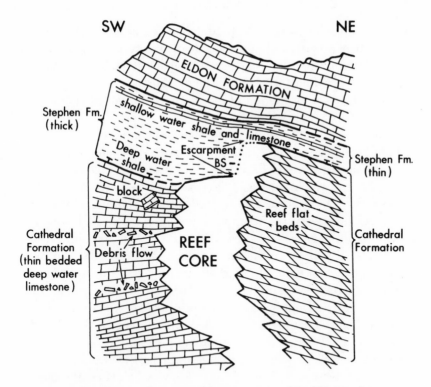

Fig. 3.2. Diagram in explanation of the exposures of Middle Cambrian
rocks on the southeast face of Mount Field shown in fig. 3.1. The reef core
in the Cathedral Formation is a coarsely crystalline dolomite that shows no
bedding. On the northeast side the core interfingers with the thin-bedded
dolomites (reef flat beds) that were originally limestones deposited in shallow waters inside the reef. To the southwest the reef core interfingers with
the muddy, carbonaceous limestones deposited in deeper waters outside the
reef. Fragments broken from the reef formed debris flows, shown interbed-
ded in these limestones. The thick Stephen Formation is above these beds;
the shale was deposited in deep water and was banked against the escarpment formed by the reef. BS indicates the approximately equivalent level to
that at which the fossils in Walcott's quarry were found. The highest part of
the Stephen Formation is shale and limestone deposited in shallow water,
and forms a thin Stephen Formation overlying the Cathedral dolomites inside the reef. Trilobites of the same age (T) were found in the basal layers of
both the thick and the thin Stephen Formation. Limestones of the Eldon
Formation overlie the Stephen rocks. Adapted from Aitken and McIlreath,
1984, *Geos*, vol. 13, p. 19, fig. 3.

dark, thin limestones were the predominant deposits. This is the transition between restricted and open seas, or alternatively from shallower, platform conditions to deeper, basinal waters. Thus a short distance west of Mount Field, at Mount Burgess (see fig. 2.1), the limestones shown on the left of fig. 3.1 have all passed laterally into shales.

The vertical wall, some 200 m high, separating dolomites of the Cathedral Formation from shale of the Stephen Formation is startling, so much so that it could have been interpreted as a fault, a vertical fracture along which movement has taken place. This cannot be so, because the Eldon Formation is undisturbed, running continuously across the area above the steep wall (see fig. 3.1). Strata below those shown in the figure are also undisturbed. The mining geologist Charles Ney, working in the area in the 1950s, observed the "west-facing precipice of dolomite" and commented that the structure was like that of a reef. Work on the assemblages of trilobites in the various layers was carried out by Fritz of the Canadian Geological Survey party. He showed that trilobites in the lowest part of the thick Stephen Formation are of the same age as those in the topmost reef flat beds of the Cathedral Formation, and in the lower part of the thin Stephen Formation overlying it (T in fig. 3.2). This was a most critical piece of evidence, showing that when the layers containing these trilobites were laid down, the wall of the submarine escarpment was as much as 160 m high.

The sedimentary rocks of the area have been studied by Aitken of the Canadian Geological Survey party, and Ian A. McIlreath. They have shown that the escarpment was part of an ancient reef structure (see fig. 3.2) that extended many kilometres northwest and southeast of the Kicking Horse River valley. The original limestone of the reef core has been altered to coarsely crystalline dolomite, destroying all the original fine structures. However, fragments that varied in size from two centimetres to several metres, broke off the reef face and slid into the deeper waters in which the Cathedral limestones were being deposited, on the slope in front of the reef (shown as debris flows and a large block in fig. 3.2). These fragments were not altered to dolomite, so that in them the branching and encrusting remains of small calcareous algae, and of branching bryozoa, were preserved. The reef thus appears to have been built by these microscopic plants and animals, an assemblage of organisms very different from the reef-frame building corals and higher algae of geologically younger reefs. The thick Stephen Formation, including the Burgess Shale, was deposited in deeper water, in successive layers that contained younger and younger suites of trilobites, and were banked up against the edge of the reef. Some limestone debris derived from the reef flat formed a thin wedge in this Stephen Formation (shown as the Boundary Limestone in fig. 3.9), and structures in the shale show that at intervals the wet muds slumped down-slope in front of the reef. Eventually deposition of mud overtopped the reef, and in these shallow waters the thin Stephen Formation was laid down over the pre-existing reef and reef flat. The Eldon Formation, of dolomites and limestones, was subsequently formed on top of the Stephen Formation and contains still younger assemblages of trilobites.

Work on the rock formations of the area, their nature, and relationships in space and time, has thus elucidated the setting in which the Burgess Shale was formed, close to a submarine cliff at the margin of the Laurentian continent. Clues had still to be sought as to why in one particular site the fossils were so exceptionally preserved. An explanation of how whole animals came to be overwhelmed and buried resulted from my initial study of a Burgess Shale fossil, the commonest animal in the Phyllopod bed, the arthropod *Marrella splendens*. The majority of the specimens, like those shown in fig. 3.3, lie parallel to the bedding—that is, parallel to the originally horizontal surface of deposition. From the short subrectangular head, two pairs of spines project, the lateral ones directed outward and curving back, the median ones curved back over the body. The head and its spines had the thickest integument, and therefore appear darkest in the photograph; the head bore the front two pairs of appendages (a_1 and a_2 in fig. 3.3). The body, with its thinner integument, tapers back behind the head, and bore limbs that had two branches. One was a jointed walking leg, the other branch consisted of a shaft bearing many filaments (these filaments may have absorbed oxygen as they moved through the water, hence this branch is referred to as the gill branch). Only a few incomplete legs are preserved on the specimen shown in fig. 3.3, but the gill branches are imbricated like slates on a roof, a thin layer of the rock matrix separating one from another and the limbs from the horns of the head. This must mean that after death the body of the animal did not settle down on the surface of the mud, to be buried as further mud was deposited on top of it. Burial in such a manner would have pressed the head, body, and limbs together into a single layer, without layers of mud separating them, gradually flattening them as more mud accumulated. The burial must have been in a manner that enabled mud to penetrate in between the delicate branches of the

Fig. 3.3. *Marrella splendens*, exposed from the under side, lying parallel to the bedding, showing an antenna, a_1, a second antenna-like appendage, a_2, lateral and median paired spines on the head-shield, l and m, incomplete walking legs, w, and gill branches, b. The dark stain behind the body is unusually large.

limbs, in between the gill branches and walking legs, and between the body and the spines of the head, before the flattening by compaction took place.

Further evidence of an unusual manner of burial is given by a small number of specimens, examples of which are shown in figs 3.4–7. In all these specimens, layers of matrix separate portions of the body, so that they lie at slightly different levels (shown by the ha-

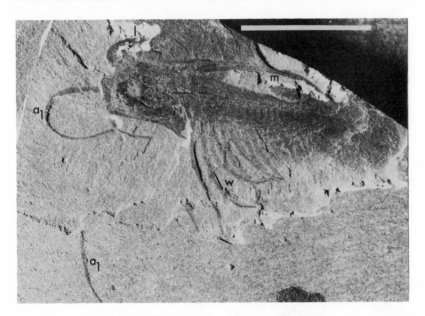

Fig. 3.4. *Marrella splendens*, lying approximately on its side relative to the bedding, labelled as fig. 3.3.

chures in the explanatory drawings of figs 3.5–7). The specimens are not seen in a symmetrical view from underneath, as in fig. 3.3, but flattened in other attitudes. For example, fig. 3.4 shows *Marrella* in side view, as a result of this individual having been buried lying on its side. In this aspect may be seen the rectangular shape of the head, a median spine curving back over the body, and a right lateral spine (partially exposed) at a lower level. The walking legs curve down below the body, and portions of the first antennae are visible. In fig. 3.5 the animal was buried at an oblique angle to the bedding, tilted down to the left and back. On flattening, the left and right spines of the head were bent differently and so are asymmetric, the front end of the head forming a blunt point and the various limbs splayed out asymmetrically. Figures 3.6 and 3.7 are of *Marrella* buried directed upward across the bedding and subsequently compacted, 3.6 inclined to the right, 3.7 tilted slightly backward. In fig. 3.6 the four spines of the head are seen at successively lower levels, and in fig. 3.7 the walking legs curve down below the body, in an imbricated series.

What was the manner in which the complete bodies of *Marrella* were buried, that allowed mud to penetrate between all the portions of the body, and resulted in most of them coming to rest parallel to

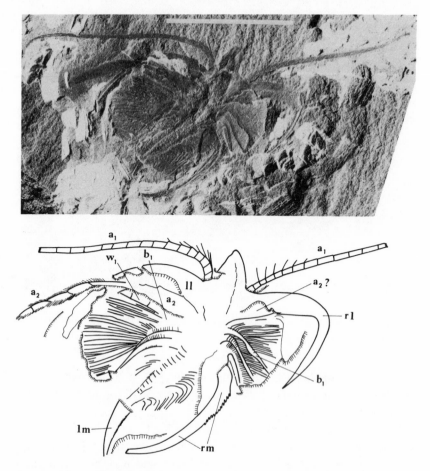

Fig. 3.5. *Marrella splendens,* an asymmetrical specimen buried at an oblique angle to the bedding. The explanatory drawing is labelled in the same manner as fig. 3.3, the prefixes r and l indicating the right and left sides of the animal. Hachures (lines with short ticks on the down-slope side) indicate a change in level. This specimen shows particularly well one of the left gills and its many fine filaments. From Whittington, 1971, *Proceedings of the North American Paleontological Convention, Chicago, 1969,* part I, p. 1184, fig. 11.

the bedding, but a few at random angles to it? It is known that at the present day, on the slight slope at the outer edge of the continental shelf, newly-deposited muds and silts, filled with water, may fail to remain stable and may consequently slump down the slope. A slight earth tremor may serve to trigger such slumps. As it moves, the sediment is thrown into suspension in a turbulent cloud. This cloud is

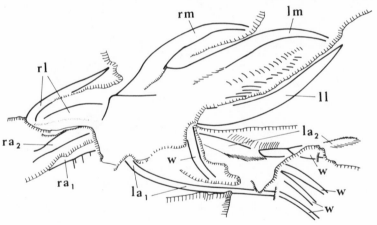

Fig. 3.6. *Marrella splendens*, buried at an oblique angle and sideways to the bedding, giving a view from the left front. The matrix has been excavated to show both pairs of spines on the head-shield. The explanatory drawing is labelled as in fig. 3.5. From Whittington, 1971, *Proceedings of the North American Paleontological Convention, Chicago, 1969*, part I, p. 1193, fig. 21.

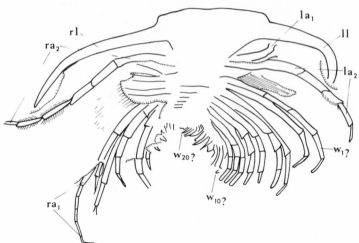

Fig. 3.7. *Marrella splendens*, buried almost at right angles to the bedding, giving an anterior view. The explanatory drawing is labelled as in fig. 3.5. The walking legs are numbered. From Whittington, 1971, *Proceedings of the North American Paleontological Convention, Chicago, 1969*, part I, p. 1192, fig. 19.

denser than the surrounding water because of the silt and mud in suspension, and so moves down-slope until the slope decreases, when the cloud slows down and the sediment settles out. Such under-water flows can also carry bigger particles along, and therefore have erosive power—Atlantic cables lying on such slopes have been cut through by them. If the muds of the Burgess Shale were deposited on such a slight slope, the animals living on and in a portion that became un-stable and subsequently slumped, would have been carried down-slope in the turbulent cloud, and buried as it settled out (fig. 3.8). As they were borne along, the mud would have penetrated between all the parts of the body, and as the cloud settled out, the bodies would have been buried, many coming to rest approximately parallel to the bedding but others at random angles. Transport may have been for only a kilometre or two (Walcott's quarry and other sites for fossils are close to the escarpment), the sediment was fine-grained, and the tur-bulence seems not to have been great enough to dismember the bod-ies. This mode of burial explains why the specimens are whole animals (virtually all known specimens of *Marrella* have the body entire and show all the limbs), and implies that these creatures were bottom-dwellers that were suddenly overwhelmed and rapidly buried. As the sediment settled out of suspension, the coarser, heavier particles came to rest first, successively finer particles later. Turbidite deposits (as they are called) thus have a sharply-defined base of coarser par-ticles, with the grains becoming progressively finer upwards from the base. Investigation of rock layers from Walcott's quarry by David J. W. Piper showed that individual thin bands (figs 2.5, 2.6) extended across the outcrop and had these characteristics, and thus supported the mode of burial suggested here.

Other arthropods have been found to be similarly preserved to *Marrella*, that is, most specimens have the exoskeleton and all the limbs, layers of matrix separate portions of the body, and only a mi-nority of the specimens are oriented obliquely to the bedding. All show the effects of compaction, the oblique specimens an asymmetry. Other fossils, such as those of worms living on or in the bottom sedi-ments, appear to have been buried in a similar manner, judging by the fact that the bodies of some specimens are twisted or folded. Rapid burial also helps to account for the many specimens of entire sponges, and one of a sponge with brachiopods attached in life position (see fig. 4.11).

Submarine slumping of sediments, as depicted in figure 3.8, ex-plains how the animals were buried. One must then ask what were

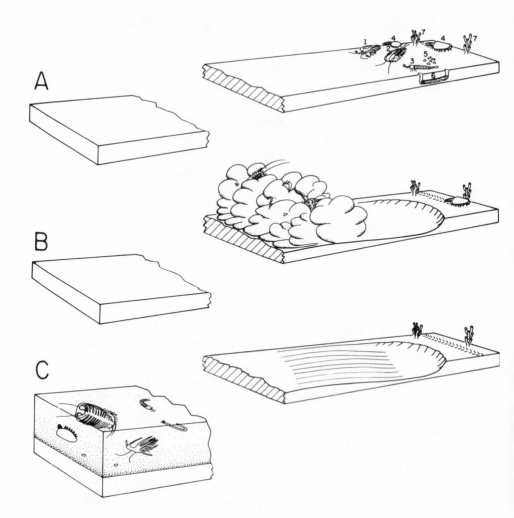

Fig. 3.8. Diagrams showing how the layers of the Phyllopod bed of the Burgess Shale, and the fossils in them, were deposited. The break in each diagram represents the unknown distance, perhaps about one kilometre or so, between where the animals lived (A, right) and where they were buried farther down-slope and consequently at a lower level (C, left). A, the pre-slide environment, shows some of the animals in the lighted bottom waters, well supplied with oxygen, in which they lived: 1–4, the arthropods *Marrella, Olenoides, Yohoia,* and *Canadaspis* respectively; 5, cap-shaped shells of *Scenella*; 6, a priapulid worm in its burrow; 7, the sponge *Vauxia.* In B, an unstable patch of mud has slumped, and has been thrown into a turbulent suspension as it moves down the slope. The animals that were living on and in the mud are carried along in the suspension, those not affected by the slump remain. On the left in C is the graded deposit made by the suspended sediment as it settled out. The deposit, filled with water, may have been some 30 to 60 cm thick. The carcasses of the animals are oriented at

the conditions in the burial site that resulted in the exceptionally per-
fect preservation of the soft parts? It would be reasonable to assume
that the sea water in the turbulent cloud, as it moved down-slope,
contained plenty of dissolved oxygen, and therefore that the layer of
wet mud in which the animals were buried would have been well
oxygenated. Under such conditions the carcasses would have been
rapidly broken down by bacterial decay and the activity of scavengers
burrowing into the mud. Examination of the turbidite layers in the
Phyllopod bed in Walcott's quarry, however, shows that this did not
happen. The layers are of even thickness of a few centimetres, are
undisturbed by burrowing animals, and show no tracks or trails made
by animals on the bedding surfaces of the layers. The environment in
which the animals were buried must have been lifeless and poor in
oxygen (anoxic). The occurrence of pyrite, in fine grains and minute
spheres scattered in the layers of the Phyllopod bed, adds to the evidence for an anoxic environment. In such environments today, for
example, within the dark, smelly muds of stagnant waters, hydrogen
sulphide is a typical product of bacterial decay. Pyrite may be formed
in such muds as a result of combination of the hydrogen sulphide with
iron. Clearly there were major differences between the lighted, oxygenated waters in which the animals lived, the pre-slide environment
as Conway Morris has called it, and the post-slide, anoxic environment. The possible locations of these environments, in relation to the
submarine cliff, are discussed below.

Study of the fossils shows that a limited amount of decay took
place in the post-slide environment, immediately after burial. For example, a conspicuous feature of the specimens of *Marrella* is a dark
stain in the rock at the ends (commonly the posterior end) of the body
(see fig. 3.3). So characteristic is this oily-looking stain that it allows
one to pick out a specimen of *Marrella* immediately the rock is split.
An analysis of the stain showed the presence of a wide variety of amino
acids, residues of the original proteins of the body. Dark stains occur
at the posterior end of other arthropod species from the Phyllopod
bed, but are far less conspicuous than those associated with *Marrella*.

These stains may have been produced by body contents seeping

---

the various angles at which they came to rest and were buried. Deposition
was in deeper water, deficient in oxygen. Traces of the activity of animals
(burrows or tracks) are not preserved in this deposit in the post-slide environment. Modified from Whittington, 1980, *Proceedings of the Geologists'
Association*, vol. 19, p. 130, fig. 2.

out of the corpse during the initial stages of decay. But decay cannot have proceeded far in the post-slide environment, because the bodies of many of the non-trilobite arthropods have the thickened cuticle of the exoskeleton and limbs preserved, as well as infillings of the gut by mineral matter or mud, and in rare cases the gut contents and traces of muscles. Internal organs are remarkably preserved in some specimens of the worm *Ottoia*. Conway Morris has interpreted different specimens as representative of the various stages of decay in the post-slide environment; in some the reflective body wall is hardly decayed at all and so conceals the internal organs (see fig. 4.19), in others decay of the body wall is advanced and the internal organs are clearly visible (see fig. 4.21). Evidently decay did not proceed far, so that the more resistant portions of the body are preserved. Conway Morris has suggested that decay in the post-slide environment may have been halted by a mineralizing solution that penetrated the Shale soon after deposition and impregnated the soft tissues. This explanation is advanced because the soft parts are not preserved as a layer of carbonaceous matter (as one might expect to be formed from the residue of decay of an organism), but as a dark, reflective layer that is a complex silicate of alumina and calcium. It is thought to be this, or some other similarly unexpected, poorly understood factor that affected the bodies after burial and resulted in the exquisite preservation exemplified by the fossils of Walcott's Phyllopod bed.

It should not be thought that all the fossils found in the Phyllopod bed are whole animals, showing only limited signs of decay, however, even if almost all the specimens of arthropods (other than trilobites) described here, and many of the worms fall in this category. There are also empty shells, showing no sign of soft parts, such as layers with abundant *Scenella* (see fig. 4.17) and the brachiopods *Lingulella* (see fig. 4.13) and *Diraphora* (see fig. 4.14), but not that of *Micromitra* (see fig. 4.12) that shows the setae. Entire, or almost entire, exoskeletons of trilobites (see figs 1.3–6) that show no appendages may be empty exoskeletons cast when the animal moulted, and carried into the post-slide environment. If so, the mechanism of transport did not disarticulate them. Eighty percent of the specimens of *Selkirkia* (see fig. 4.23) are empty tubes, some of which were empty in the pre-slide environment and acted as a base to which other animals living there attached themselves (see fig. 4.64). About half the specimens of the trilobite *Naraoia* (see fig. 4.30) and the crustacean *Canadaspis* (see fig. 4.32) are without soft parts. Whether these are moults of the exoskeleton, or the remains of carcasses after the soft parts decayed in the pre-slide environment, is uncertain. Also difficult to explain are the clusters of *Canadaspis* (see fig. 4.32) of up to a hundred carapaces with and without soft parts. They may be a mixture of moults and whole animals brought together in the course of transport to the post-slide environment. Thus the Phyllopod bed is like most fossil-bearing beds in that it contains the remains of animals after decay, the hard parts. Its unusual feature is that great numbers of live, whole animals were carried into the site of deposition, and were preserved in a most exceptional way.

While many of the factors that determined the manner of preservation must have been operative very soon after burial (that is, within a few months or years, perhaps), a longer-term (hundreds or thousands of years) factor was the compaction (referred to above) of each layer of sediment as it was buried by subsequent layers. The water in the layer of mud was squeezed out and the mineral grains and flakes packed together, so that the layer of rock formed was about one-tenth, or even as little as one-twentieth, of the original thickness of the wet mud. This compaction of the mineral flakes imparted a fissility to the rock, so that it splits parallel to the bedding. In a layer of mud containing animal bodies, the bodies were flattened in the same proportion as the matrix, and mud between regions of the body was reduced to a thin layer of this matrix. In the case of arthropods buried parallel to the bedding, the originally convex exoskeleton was flattened and the regions of the body brought close together, as in *Marrella* (see fig. 3.3) or *Molaria* (see fig. 4.43). Specimens which were buried lying on one side relative to the bedding, when flattened (*Marrella*, see fig. 3.4; *Molaria*, see fig. 4.44), preserve the original convexity and are most valuable in making reconstructions. Bodies buried at an oblique angle to the bedding were rendered asymmetrical by this flattening (compare fig. 3.3 with figs 3.5 and 3.6). The branches of the limbs of these arthropods are arranged in an imbricated layer just beneath the exoskeleton in many examples (*Marrella*, see fig. 3.3; *Olenoides*, see figs 4.26–28; *Burgessia*, see fig. 4.56), the walking legs curved backward or forward. This is not a natural position of the legs, for in life they would be curved downwards. This re-arrangement of the limbs in relation to the body may have been brought about during transport. If decay of muscles and ligaments had occurred in the carcass before burial, in the pre-slide environment, the limbs could have been swung backwards or forwards, by the turbulence, in which case compaction after burial would have brought the limbs still closer to each other in an imbricated series and close under the body, as the thickness of the layer of sediment separating them was reduced in compaction. Thus the appearance of the fossil as preserved—for example, in the arthropods and the polychaete worms with their imbricated bunches of setae (see fig. 4.24)—is influenced by compaction as well as other factors.

This lengthy discussion of how the animals and shells were buried, and what happened afterwards, arises from comparing and contrasting individual specimens of a species, such as those of *Marrella* or *Ottoia*, and attempting to explain similarities and differences. The unusual mode of burial deduced from such attempts implied that parts of the body of an animal were buried at different levels in the rock. When a specimen is revealed by splitting a piece of rock from the Phyllopod bed, this split will tend to pass along larger, flattened portions of the body (such as the head and its spines of *Marrella*) and leave other portions, for example, the limbs, buried in the rock. As explained in chapter 4, such buried portions of the body may be revealed by removing the rock matrix, and much new information about the fossils has been gained in this way. Comparisons of specimens have shown that some decay occurred, but that it did not proceed far, for many specimens are of whole animals. Post-burial events were complex and are by no means fully understood.

To conclude this chapter, the results of the investigations of the Shale and the rocks adjacent to it, and of the fossils, are brought together to give a picture of the site of deposition. Figure 3.9 shows the submarine escarpment of the thick dolomite of the Cathedral Formation, which varied in height from 100 to 300 m along its length, and the muds of the Stephen Formation that were banked in cusps against it. No canyons breaching the wall of the escarpment are known, but at times fragments of limestone were washed out over the edge from the shallow waters, forming, for example, the wedge of Boundary Limestone jutting out into the Stephen Formation. The figure shows a typical stage in the accumulation of the Burgess Shale portion of the Stephen Formation. Wet muds on the slopes of the cusps were unstable, and patches slumped down-slope from time to time. In the deeper, quiet waters between two cusps, close to the bank, oxygen may have been used up, resulting in an anoxic environment in the mud and water immediately above it. A slump parallel to the face of the cliff, of mud and the animals on and in it, into such a spot, may have given the peculiar conditions that resulted in exceptional preservation. Slumps causing burials in other places where oxygen was not used up would not provide special preservation. Such preservation

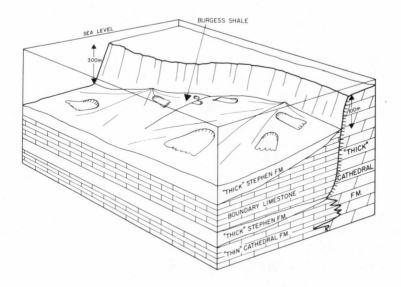

SEA LEVEL

300m

BURGESS SHALE

100m

"THICK"

CATHEDRAL

F.M.

"THICK" STEPHEN F.M.

BOUNDARY LIMESTONE

"THICK" STEPHEN F.M.

"THIN" CATHEDRAL F.M.

Fig. 3.9. Block diagram to show the environment in which the Phyllopod bed of the Burgess Shale was deposited. The vertical wall (see figs 3.1, 3.2) between dolomites of the 'thick' Cathedral Formation and the 'thick' Stephen Formation is shown as a submarine escarpment that overlooked a basin being filled with muds of the Stephen Formation. The scars of slumps that have occurred in the muds are shown. Some resulted in deposits in a depression close to the escarpment, and formed the Phyllopod bed. The Boundary Limestone is a wedge in the Stephen Formation, formed earlier when fragments of limestone were swept over the edge of the escarpment into the basin. From Whittington, 1980, *Proceedings of the Geologists' Association*, vol. 91, p. 136, fig. 4.

was thus a very localised, unusual event, that was repeated only rarely during a short period of time; as a result the finest fossils do not occur throughout the 2.3-m thickness of the Phyllopod bed, but are confined to four or five thin layers within it. The area over which the fossils found in Walcott's quarry occur is unknown. They have been found only along some 60 m of the width of the quarry face. To the north the escarpment wall is close by, while to the south of the quarry the strata are folded adjacent to a vertical fault along which there has been considerable movement. The geological structure of the area in which the Burgess fossils lie is complex, the result of great earth movements in Mesozoic time. Thus, for example, the Wapta Mountain–Mount Field area is part of a larger mass of rocks that have been transported eastwards over 160 km on a gently-inclined surface, a thrust fault. A second such fault underlies the higher portion of Wapta Mountain. These earth movements brought the thick Stephen Formation (including the Burgess Shale) to a site where it was not subjected to great heat or pressure which might have destroyed the fossils. During these earth movements the shales, being close to the relatively rigid wall of the escarpment, were protected from stress and not distorted. Thus the preservation of the Burgess Shale fossils is a result of the combination of rare conditions of deposition close to a massive wall, and the fortune of these rocks having been carried by earth movements to a site where they were not altered later. Subsequent erosion of the transported piles of rocks has produced the mountains of today and exposed the shales on the ridge.

It has been mentioned above that the submarine escarpment has been traced by Aitken and McIlreath not only north from the Burgess Shale section, but some 20 km southeast, beyond the Kicking Horse River valley. Conditions of deposition of shales banked against this wall were similar along its length, so it was to be expected that other localities in which fossils were preserved in like manner to those of the Burgess Shale would be found. In their explorations in 1981 and 1982, Collins, Briggs, and Conway Morris found more than a dozen such sites, all close to the escarpment, some on the southeast face of Mount Field, others on Mount Stephen and other places to the southeast. They are at a similar stratigraphical level to the Walcott quarry, or slightly higher or lower. The soft-bodied fossils include a wide variety of the species found by Walcott, similarly preserved, and some that are rare in his collection are quite abundant. No new kinds, or specimens better preserved than Walcott's, have yet been found, however. Walcott's discovery and the hard work he put in to exploiting it to obtain his great collection, have provided a rich treasure house of material for investigation, and the foundation for the new studies of the fossils described here.

4 *Fossils of the Burgess Shale*

With very few exceptions the fossils from the Burgess Shale that are illustrated herein (scale bar 1 cm, unless specified otherwise) are from the 2.3-m-thick Phyllopod bed in which Walcott excavated his quarry. The vast majority of Walcott's specimens are from this bed, and these fossils are better preserved than any from the Raymond quarry or the other localities mentioned in chapter 2. The discussion in chapter 3 of the environment in which the Shale was deposited and of how the fossils were preserved, refers primarily to the Phyllopod bed.

Each kind of fossil has a name, composed in a particular way, and accepted from the earliest published description. These names are formed in the same way as those for living animals and plants, fossil and living organisms being classified in a single system. Before embarking on descriptions of the Burgess Shale fossils, I discuss these matters and some of the special methods used in studying them. A list of species is given in the Appendix, followed by a list of publications by Walcott and other authors.

NAMING AND CLASSIFYING FOSSILS

The confusion that arises in using the same common name for different things—'corn' is maize in the United States, but wheat in England—is well known, as is the use of different common names for the same thing. In natural science it is essential to have one name for each entity, a name used and understood by all scientists in all countries.

41

The Swedish naturalist Carolus Linnaeus proposed the system we use today in the tenth edition of his *Systema Naturae* (1758). Each species of animal or plant has a one-word name, its specific name, which must be Latin or latinized. Originally names were formed from Greek or Latin roots describing some character of the animal or plant; this practice continues today, but latinized surnames, place names, or arbitrary combinations of letters are also used. Linnaeus grouped similar species, which were presumably related to one another, into a *genus* (plural, *genera*). Generic names are made in the same manner as specific names, and are descriptive names referring to some distinctive character of the group of species, or of the particular species which the author of the genus considers characteristic of it. When referring to a species the generic name is used in combination with the specific name.

Linnaeus extended his hierarchical system by grouping genera into families, families into orders and classes, the latter being main subdivisions of a major division of animals, a *phylum* (plural, *phyla*). Names of these higher categories may be descriptive of a distinctive character of the group, but particularly in the case of a family may be based on the name of the genus selected as the type of the family. The names of phyla of animals are used as headings subdividing this chapter. There are recognised terminations for such names, for example, Arthropoda, terminating in 'a', for animals with jointed legs, but this group may be referred to less formally as arthropods. Similarly, the term 'agnostoid' is used in the caption of fig. 1.5, referring to Agnostoidea, one of the orders of the Trilobita. At the first mention in the text, and in the captions to the figures, the generic and specific name is given. Thereafter in the text the generic name only is used, as a convenient shorthand, to refer to the particular species named.

Over one hundred species of animals and plants are known from the Burgess Shale (see Appendix). Walcott's collection contains one or two, or tens, or hundreds of specimens of each species. These specimens provide a sample of the individual animals that constituted the original living population. Each of these samples affords a basis on which the variation between individuals of the species may be described and characterized. The sample of specimens may show a range in size, or a variation between individuals in the relative proportions of particular parts of the body. If these variations cluster about a mean, one may assume that the sample is of individuals of the same species. In the case of the species that have the soft parts preserved, many characters of the animal are available for study, for example, the num-

ber, type, and position of appendages; consequently there is no difficulty in separating specimens of one species from those of all other
species. Difficulties arise, however, when hard parts only remain, in
identifying, for example, isolated carapaces of *Canadaspis* (see fig.
4.32) in relation to carapaces of which the body and appendages are
known (see figs 4.33, 4.34). Only specimens showing features in addition to the carapace may be identified with certainty as *Canadaspis
perfecta*. Other genera of phyllocarids, represented by more than one
species (see the list of species), have been erected solely on the basis
of the carapaces, because soft parts are unknown. Some, such as
Isoxys and *Proboscicaris*, are recognised by the distinctive shape, *Tuzoia* by the external ornament. Species are much more difficult to
characterize. How to define a species—how great a range of variation
in form to include in a particular species, and how great a range in
time and space to allow—is a problem familiar to all palaeontologists
dealing with hard parts only. An additional question is whether the
species of the palaeontologist is equivalent to one recognised among
living animals.

In considering the Burgess Shale animals with soft parts, one may
feel reasonably confident that the specimens are derived from the successive populations of the species that lived in the immediate area
during the span of time in which the Phyllopod bed was deposited.
They were thus derived from interbreeding populations related to one
another through time and seem to be species in the same sense as
those recognised among living animals. In attempting to define species in palaeontology, only experience in dealing with fossils will lead
to judgements that may prove widely acceptable. In the assignment
of a species to a particular genus, and of a genus to a particular family—experience and judgement are paramount.

In formal scientific writing, as in the list of species from the Shale
given in the Appendix, the name of the author who originally described the species, and the date of the publication in which it was
first made known, are given as a brief reference to the origin of the
name. Thus *Oryctocephalus reynoldsi* Reed, 1899 (fig. 1.6) is so
named because F. R. C. Reed described this species in the *Geological
Magazine* published in 1899, naming it for S. H. Reynolds who collected specimens on a visit to the fossiliferous locality discovered by
Klotz on Mount Stephen (see fig. 2.1). The final resort in identifying
another specimen, collected later from Mount Stephen, or indeed
from Walcott's quarry, is comparison with the actual specimens described by Reed, which are in the Sedgwick Museum, Cambridge. I

draw attention to this to emphasise that in the identification of a fossil, not only is detailed study of exactly what it is like essential, but also thorough study of the relevant writings and earlier collections. Only then may conclusions be drawn on similarity to, or differences from, other like fossils, and thus on evolutionary relationships, and distribution in space and time. An adequate library and access to museum collections are obvious requirements. In the end, of course, whether a fossil species is new to science, or the same as one already known, is a matter of judgement and experience. The complications are considerable. How can one be sure that no one in any publication in any country has described a fossil that is sufficiently similar to the one you have just collected to be regarded as the same species? The agnostoid trilobite shown in fig. 1.5 was originally described as a new species, *Triplagnostus burgessensis*, known only from Walcott's quarry. But I name it *Ptychagnostus praecurrens*, accepting the judgement of an expert, Richard A. Robison, of the University of Kansas, who recently gave his reasons for believing that these Canadian specimens are identical with some from Sweden described earlier as *Ptychagnostus praecurrens*. In such a case, the earliest published name for the species must be used, under international rules agreed upon and accepted by palaeontologists and zoologists. In Robison's opinion the species *praecurrens* belongs within the genus *Ptychagnostus*, the earliest generic name proposed for species of this type. The Swedish palaeontologist Westergaard who first described the species placed it in a different genus. That the generic assignment of the species *praecurrens* has changed is indicated by placing the author's name and date in parentheses, as in the Appendix.

NAMING AND STUDYING THE BURGESS SHALE FOSSILS

As noted in the example above, problems may arise in identifying as accurately as possible the Burgess Shale fossils, not only in dealing with trilobites, but also with brachiopods or with other groups which have mineralized shells and are known from Cambrian rocks in many countries. In dealing with the species of soft-bodied forms, the problem is not one of comparing them with earlier-named species. The preservation reveals much more detail than is usual in the body of the fossil, and most species of worms, arthropods, or other large groups are strikingly different one from another. They are hard to compare with nearly all other fossils, even those from other strata that show exceptional preservation, or with living animals. The problem is to discern their evolutionary relationships and to give each one an appropriate place in a classification based on these relationships. It is most extreme in dealing with the species included under Miscellaneous Animals, which do not fit into the highest (and therefore most widely-conceived) place, the rank of phylum. Some similar problems in classification within the highest rank, illustrated by the various arthropods, are discussed below. Broader questions of evolution and relationships are considered in the last chapter.

Despite the exquisite preservation, small individuals of any species (less than two or three millimetres in length) are very rare, although the surfaces of the Shale have been searched with a binocular microscope. Still smaller specimens have not been found, so that no evidence of the earliest growth stages is available for any species. Growth stages of related species may be similiar, and their absence means that a potential source of clues to relationships is not available.

Microscopic single-celled organisms have been extracted with acid (because they had an organic covering that is acid-resistant) from Precambrian rocks, so they must have been abundant in Cambrian times. They may be present in the Burgess Shale, but no evidence has been reported.

In describing the many new kinds of animals, Walcott had to create new names for them, for both the genus and the species. For genera, he used as roots the names of mountains (for example, *Waptia* from Wapta Mountain), lakes (*Emeraldella* from Emerald Lake, visible from his quarry), or other topographical features in the neighbourhood. Occasionally he used personal names, such as *Marrella* for his friend John E. Marr of Cambridge University or *Vauxia* from his wife's maiden name, Vaux. Names had been given to topographical features by early surveyors (such as Mount Burgess by Klotz) or explorers. One of the latter was a Yale University graduate, Samuel E. S. Allen, who had climbed in Europe before becoming one of a small group (the so-called Yale Lake Louise Club) that in 1891 to 1894 explored the region around Lake Louise. From their data they prepared the first map of the region, on which Allen, after consulting local Indians, gave particular features such Indian names as Opabin (rocky), Odaray (cone-shaped), or Wiwaxy (windy). From these were derived some of the more exotic names that Walcott gave to his new genera of fossils. Specific names may be either descriptive (for example, *gracilens, splendens*) or again in honour of a particular person (*brocki* for T. E. Brock, then Director of the Geological Survey of Canada). New genera resulting from recent studies have been named descriptively, for example, Conway Morris's *Hallucigenia* refers to the "bizarre and dream-like appearance of the animal" (to quote the author), and his *Dinomischus* is derived from the Greek for goblet (*dinos*) and stalk or stem (*mischos*).

Walcott's descriptions of the Burgess Shale fossils were short; they were illustrated by photographs and gave a brief account of how each species was related to other species of fossils, and how it fitted into classifications of living and fossil forms. He made over sixty new genera and many new families, because these fossils were so different from any others known, but then he fitted these groups into higher categories (orders and classes) already established. His work was, as he himself recognised, preliminary, and only in a paper published after his death in 1927 (based on manuscripts he left) were reconstructions of the species given. At about the same time, G. Evelyn Hutchinson, Rudolf Ruedemann, and Charles E. Resser described other specimens in Walcott's collection, and Raymond reported on the new material he had collected. Leif Størmer came from the University of Oslo as a young man to work with Raymond, and made an important re-assessment of the arthropods based on his examination of specimens illustrated by Walcott, his study of Walcott's writings, and comments by subsequent authors. Walcott had, however, done much more work than his publications indicate. The pieces of shale containing each of the approximately sixty thousand specimens in his collection had been neatly sawn into rectangles, and all those of each species Walcott described had been brought together for more detailed study. Many new, undescribed forms had been recognised by Walcott during this sorting, some of which had been photographed and given labels bearing suggested new names.

This great store of exquisite material was dipped into by an Italian zoologist, Alberto M. Simonetta, who examined the arthropods and a few of the strange organisms. Since 1962, he and his collaborator, Laura Delle Cave, have published a number of articles containing the first restorations of these animals, as well as descriptions of new species and of ten new genera, many based on the undescribed forms sorted by Walcott. Their restorations are attractive-looking, but the photographs accompanying them do not show many of the features described. Further, the specimens were not prepared in any way but were photographed just as Walcott left them. Despite these recent studies, therefore, much remained to be done: the specimens needed to be examined in detail to see if more could be revealed, and it was essential that better photographs be obtained.

When the shale is split along a bedding plane and reveals a fossil, this split may leave some portions of the fossil still concealed in the rock. Further, some portions may be on one side of the split, some on the other (these two sides are referred to as part and counterpart; for an example, see fig. 4.58), because of the way the fossils were buried, with rock matrix between different portions of the body. It is thus important to keep part and counterpart together, and necessary to remove some of the rock matrix in the laboratory, to expose the fossil as completely as possible. This is a delicate and time-consuming task—laboratory work on fossils takes far longer than collecting them in the field! Far better binocular microscopes and lights are available today than were available to Walcott, and delicate tools unknown to him. The specimens show that Walcott used a small hand chisel and hammer to prepare some specimens in a crude way, whereas my colleagues and I prepared specimens under the microscope, using as a minute chisel a sharpened needle driven by a small electric motor of the type employed in a dental machine. This attachment converts the rotary action into an up-and-down movement, giving a microscopic version of a road drill. The matrix will part from the surface of the fossil fairly readily, so that, for example, if one digs down cautiously beside the margins of a concealed spine or limb, when the right level is reached the matrix may part and lift off the fossil. Such excavations appear as light or dark (in shadow) places in the photographs, and small pits show the inevitable accidental damage to the fossil. For example, in fig. 3.6 there are small white marks left by the micro-drill beside the lateral and median spines and the left antenna. By contrast, the chisel-marks of Walcott's crude preparation on the left and lower sides of fig. 3.5 are broad. Figure 4.58 shows a part and counterpart and how concealed limbs were exposed in the part by cutting through the body to the appropriate level.

The next task was to obtain good negatives that could be enlarged to reveal the details of the specimens. Experiments showed that photographs taken in ultra-violet radiation gave exceptionally sharp negatives and enhanced the contrast between fossil and matrix. Almost all the photographs were taken this way. As already explained, the fossils may not be in one plane in the rock, but in several, and the relative levels of portions of the animals reveal their relationships. For example, in fig. 3.6, the left lateral, left median, right median, and right lateral spines of the head lie at successively lower levels, and their relationship to one another is clear. A photograph taken with the radiation directed at a low angle over the specimen brings out these different levels, as in fig. 4.28. These show that the walking legs are nearest the observer, the gill branch below, and the exoskeleton concealed still deeper. The animal is therefore being viewed from below, and in life the gill branches were extended back above the legs. By contrast, when the radiation is directed at the specimen at a high angle, it will be reflected back from the shiny film (see fig. 4.26) and give a different result, with the changes of level less apparent but details brought out remarkably well (for example, figs 4.2, 4.3). Fig. 4.16 shows the same specimen photographed in these two different ways. Immersing the specimen in water or alcohol reduces the scattering of the radiation and may enhance the reflectivity (see fig. 4.27), and bring out details by increasing the contrast between different portions of the specimen and between the specimen and the matrix (see figs 4.49, 4.69).

However clear a photograph, unless it is labelled (as in fig. 3.3) or explained, the reader unfamiliar with the particular fossil may not be able to appreciate the evidence for a particular structure or judge its validity. This is especially critical when dealing with ancient animals, which may well have structures unparallelled in fossils of lesser age or in living animals. As a way of explaining how a particular specimen is interpreted, a labelled drawing accompanies the photograph (for example, figs 4.58, 4.59). Such drawings are made using a camera lucida, a device that reflects the image from one tube of a binocular microscope on to a piece of paper beside it. One can thus look at a specimen through the microscope and draw it at the same time. Since the drawing is labelled, the accompanying photographs may be left untouched. To make a drawing one must study the specimen in detail and decide how to interpret what one sees—a salutary exercise that has given rise to an enormous amount of new information, thus improving the understanding of these fossils. This is not to say that all the structures revealed by the fossils are fully understood, for some remain enigmatic. The final stage is reconstruction (for example, fig. 4.61)—trying to portray the animal in its original three dimensions in a life attitude. One of my colleagues in this work, Bruton of the University of Oslo, has gone a stage further with the able technical help of Aage Jensen of the Museum of Palaeontology. They have made models (figs 4.40, 4.50) that have been cast and hand-carved in plastic, which make an elegant final touch to a museum display.

These methods of study have been used over the past fifteen years by Bruton and me, and by my colleague at Cambridge University, Hughes, and two former research students, Briggs and Conway Morris. The first four of us mentioned have been studying the arthropod species, while Conway Morris has tackled the worms and sundry other strange animals. Donna L. Satterthwait, while a research student at the University of California, partially re-studied the algae, but only a brief abstract of her results has been published. The large variety of sponges is currently being studied by Rigby of Brigham Young University, Provo, Utah. The following account draws on published work, and hence inevitably concentrates on arthropods, worms, and particular miscellaneous animals. The relative proportions of different groups of animals in the fauna are shown in a bar graph in fig. 5.2. For convenience, the name 'algae' and the names of each of the major animal groups have been used to subdivide the following discussion.

ALGAE

Much the commonest species of alga in Walcott's collection is represented by the irregular fragments of perforated sheets that he called *Morania confluens* (fig. 4.1). Each piece of rock containing such fragments is crowded with them, but no other alga accompanies them and only a few animals; these include the polychaete worm *Burgessochaeta* and rare examples of *Marrella* and *Burgessia*. The delicate, branching *Marpolia spissa* (fig. 4.2) occurs also as masses of broken fragments, not associated with *Morania confluens* or any other alga. One slab contained *Marpolia* and the animals *Eldonia* and *Wiwaxia*. Walcott described several additional species of *Morania* and other genera and species of branching forms (fig. 4.3) but all are quite rare in his collection.

The algae are all fragmentary, and are preserved in the form of thin, shiny films. By covering a specimen of *Marpolia* with a film of transparent balsam and then peeling it off, the English palaeobotanist J. Walton removed the thin layer in which the alga is preserved. Under the microscope individual branches showed darker longitudinal strips, others transverse dark bands, which he concluded were evidence of cell structures. No similar technique has been used to investigate other specimens of the algae, nor has any re-study of them been published. The way the two commonest algae occur, almost in isolation from one another and the fauna, is difficult to explain. Perhaps the pre-slide environment in which they lived was different from that of most of the animals, for slabs of shale which show a variety of

animals do not seem to contain algae. Or perhaps the algae lived attached to the submarine cliff at a shallower depth, and were periodi-
cally broken off and swept down to burial in separate slumps.

PORIFERA

Porifera (sponges) have a conical-cylindrical, vase-shaped or globular
body that may be branched. The walls are traversed by passages lined
with cells having minute, whip-like processes; the movement of these
processes draws water through the walls and forces it into the central
cavity and out through a central opening. The cells of the sponge ex-
tract nutrients from the water as it passes through the walls. These
walls are supported by an internal framework of spicules, made of
either calcium carbonate or opaline silica; spicules vary in form and
may be cemented together to produce a rigid framework. Other
sponges have a skeleton of tough, flexible organic material, called
spongin, and are familiar as commercial sponges.

In numbers of genera (see fig. 5.2) sponges are the second most
varied group in the Shale, and they predominated among animals
fixed to the sea bottom. If specimens are counted, *Vauxia gracilenta*
(fig. 4.4) is by far the most abundant in Walcott's collection. Rigby,
who is studying the Burgess Shale sponges, regards *Vauxia* as having
had a skeleton of spongin; the enlargement in fig. 4.4 shows the complex network, including strands that formed a polygonal network
around openings in the wall. An example of a sponge that had a thin
wall of siliceous spicules is the elongate *Leptomitus lineata* (fig. 4.5).
The outer layer of the wall consisted of two different sizes of vertical
spicules, the inner layer of transverse spicules. Species of this genus
are known in older Cambrian rocks, and it may be that kinds with a
more complicated skeleton are derived from the relatively simple
Leptomitus. An example may be the fairly common *Pirania muricata*
(figs 4.6, 4.7), which had a central hollow stem to which tufts of spicules were attached, directed upwards and outwards. *Choia ridleyi*
(fig. 4.8) was rarer, and instead of being anchored at the base to the
sea bottom, like *Pirania*, its circular body was propped in position by
long spicules projecting radially from it. Peculiar in structure, and of
unknown relationships (like so many of the soft-bodied fossils from the
Shale) is the rare *Takakkawia lineata* (fig. 4.10), in which the tracts of
spicules that ran up the wall were twisted in a spiral. A different class

of sponges is represented by *Protospongia hicksi* (fig. 4.9), with its geometrical pattern of spicules, a type that is widely known from Cambrian rocks.

BRACHIOPODA

In brachiopods (briefly discussed in chapter 1) the shell is lined by a sheet of tissue, called the mantle, which encloses the body. A pair of muscular arms bearing tentacles (the lophophore) enables the animal to create water currents that bring in microscopic particles of food and oxygen, and carry out waste products. Brachiopods are held, in a position that keeps the opening between the valves clear of sediment, in most cases by a muscular stalk (the pedicle) attached to the sea floor, a hard rock surface, or an empty shell. Inarticulate brachiopods were abundant among Cambrian fossils; one example from the Burgess Shale, of *Micromitra burgessensis* (fig. 4.11) shows several shells that were originally attached by the pedicle to spicules of the sponge *Pirania*. The two species of animals were associated in life, attachment to the sponge keeping the brachiopods off the sea floor, in a favourable position for feeding (which was perhaps aided by the feeding currents of the sponge). It is very rarely indeed that the fine hair-like processes (setae) that fringed the mantle are preserved, but these may be seen around the margins of the shells in fig. 4.11, and more clearly in the isolated example shown in fig. 4.12. A second kind of inarticulate brachiopod, *Lingulella waptaensis* (fig. 4.13) is tongue-shaped; shells of this kind are common in Cambrian and Ordovician rocks. Articulate brachiopods, the two valves of which are linked by a pair of teeth and sockets, are much more abundant in post-Cambrian than Cambrian rocks, and are still numerous and of varied kinds in today's seas. One of two species described by Walcott, *Diraphora bellicostata* (fig. 4.14), shows no trace of soft parts, and was presumably an empty shell, the two valves still linked together. Because of geological events since burial, the calcium carbonate of the shell was dissolved, and only the impression remains.

LOPHOPHORATA

The name 'Lophophorata' has been used for a super-phylum embracing animals having a lophophore, for example, brachiopods and bryozoans, and this group is a convenient one in which to place the unique animal *Odontogriphus omalus*. The single known specimen, preserved as part and counterpart, appears merely as an irregularly-bounded, darker patch on the shale, roughly oval in outline, about 6 cm long, and crossed by various fractures. It seems to represent a gelatinous body, that is traversed by closely-spaced, parallel, dark lines which suggest that it was annulated. At one end is a minute (4 mm across) horseshoe-shaped structure with a darker, median band behind it. In this horseshoe-shaped structure are preserved the impressions of extremely small (0.4 mm long), tooth-like objects, each having a broad base and a long, thin cusp. Conway Morris interpreted the horseshoe-shaped structure as the two tentacle-bearing arms of a lophophore, and the tooth-like objects as internal supports for the tentacles. This feeding apparatus surrounded the mouth at the front end of the body, the dark strip behind it the trace of the gut. As reconstructed (fig. 4.15), the animal (*Odontogriphus* means "toothed riddle", and *omalus* refers to the flattened form) probably floated, or swam feebly by undulating its flattened body, in surface waters. The supposed tentacles were probably covered by fine, hair-like processes that created currents to carry tiny food particles into the mouth.

The interest of this unpromising-looking specimen is not only its rarity, but the fact that it is the remains of an animal that bore minute, tooth-like objects. Such objects have long been known from rocks of late Cambrian to Triassic age, and are abundant and varied in structure in the post-Cambrian periods. They are phosphatic in composition, and so may be extracted from the matrix by dissolving it in acetic acid. They are thus isolated from any remains of the animal they belonged to. Conodonts, as these tooth-like fossils are called, have been the subject of much speculation as to their function and zoological affinities. The conodont-like objects in *Odontogriphus* are only impressions, so their composition cannot be determined. Hence it cannot be established that *Odontogriphus* is related to the animal that bore post-Cambrian conodonts. Such an animal has recently been found in Carboniferous rocks. The elongate body had at the front end an assemblage of the more complex, geologically younger types of conodonts, which probably functioned as a food-grasping apparatus. It was quite different from *Odontogriphus*; thus there were two distinct, unrelated kinds of animals that had internal tooth-like structures. The problem of what the conodont-bearing animal or animals were like illustrates the difficulties of dealing with common but isolated parts of animals preserved as fossils. Rare but exceptionally-preserved fossils are of prime importance in the solution of such problems.

COELENTERATA

This major group of animals includes sea-anemones, corals, and jelly-fish, all of which have a relatively simple anatomy, a cup- or umbrella-shaped soft body with a central digestive cavity, and a mouth surrounded by tentacles. Characteristic are the cells in the tentacles which carry a hollow, thread-like tube that injects prey with poison, as the tentacles draw it to the mouth. Fossils of corals, which secreted a calcium carbonate exoskeleton, are common and varied from the Ordovician period onwards. Remains of such soft-bodied coelenterates as sea-anemones and jelly-fish rarely are found as fossils, but supposed jelly-fish and large branching forms regarded as coelenterates are abundant in certain late Precambrian rocks. Thus one might expect fossils of coelenterates in the Burgess Shale, but Walcott described only one species that he considered to be jelly-fish, and one that was possibly a branch of coral. The former (*Peytoia*, see under Miscellaneous) is part of another animal, the latter (*Margaretia*) is thought to have been an alga. However, *Mackenzia costalis* (fig. 4.16), described in 1911 by Walcott as a sea-cucumber, is now considered to be a sea-anemone. The ridges on the elongate body are interpreted as the ends of radial partitions. Walcott showed a ring of plates around the mouth at the upper end, but these plates disappeared after the specimen was washed in acid. Whether they were part of the specimen or a deposit on the surface of the fossil is uncertain; they are not there now, and how this fossil is to be interpreted is problematic.

Simonetta and Delle Cave illustrated and briefly described the part and counterpart of a large specimen as *Fasciculus vesanus*, and considered it to be possibly a cnidarian. The term 'Cnidaria' is used as an alternative to 'Coelenterata', or by some authorities to include both coelenterates and ctenophores. The latter are globular in shape, lack stinging cells or hard parts, swim or drift in the sea, and have cilia (hair-like processes) in bands on the body. Collins, Briggs, and Conway Morris illustrate a small additional specimen of *Fasciculus* from a locality on Mount Stephen, and suggest that it may be a ctenophore, the group being otherwise unknown as fossils.

Coelenterates are thus not an important portion of the Burgess

Shale fauna (see fig. 5.2), but whether this reflects their original rarity
in the pre-slide environment, or whether they were present but for
some reason not preserved, is uncertain.

MOLLUSCA

As explained in chapter 1, during the Cambrian period most shells
that may be referred to the Mollusca are unlike the common molluscan shells of the post-Cambrian periods. One shell common in the
coarser, silty layers is the small, cap-shaped *Scenella* (fig. 4.17). Another is *Hyolithes carinatus*, elongate, tapering, trigonal in cross-
section, having in the best-preserved specimens a small lid at the ap-
erture (fig. 4.18). From each side, between the shell and the lid, a
curved blade-like plate extends. The lower side of the cone was gently
curved, and these lateral blades helped to prop the animal, known as
a hyolithid, in position on the sea floor. Presumably the lid, when
closed, served to protect the animal, but could be opened for feeding
and obtaining oxygen. The relationships of this animal have been
much debated—should it be regarded as a mollusc or as a representative of a separate group? Whatever they were, it is apparent that the
small hyolithids provided food for larger predators, because the shells
have been recognised in the gut contents of the worm *Ottoia* and the
arthropod *Sidneyia*.

WORMS

Long, narrow, silvery impressions in the Shale, some showing a pair
of tentacles at the front, others having bunches of bristles along the
body, or the body covered by scales, were described by Walcott as
annelid worms. They were then, and still remain, the best-preserved
fossil worms known. Their variety shows how early such animals had
diversified in the marine realm, and provides critical evidence of their
evolution. Walcott classified them into several groups of worms known
living today, but the relationships of particular species were immediately challenged and have continued to give rise to debate. Conway
Morris has been engaged in the first detailed examination of Walcott's
large collection and has produced a wealth of new information and
interpretation. As shown in succeeding sections, various forms, such
as *Wiwaxia* and *Pikaia*, are not worms, and many new kinds of animals
have been distinguished.

But of course Walcott was correct in thinking that there are many species of fossil worms, many more than his preliminary studies recognised. His collection contains over a thousand specimens of *Ottoia prolifica* (figs 4.19–22), its large body annulated, and at the front end a proboscis with many tiny hooks and spines. This proboscis could be extended by means of muscles in the body wall that compressed the fluid-filled body cavity, and could be retracted by other muscles. It was used in burrowing and in capturing prey. At the posterior end are a group of hooks. The animal was thus well adapted for burrowing in search of food, the hooks of the proboscis and the posterior end being used to hold the body in place as it extended and contracted in the burrowing process. Particular specimens of *Ottoia* show remarkable details of not only the proboscis, but also the gut, muscles, and nerve chord, so that Conway Morris was able to draw a hypothetical dissection of the animal (fig. 4.22), such as one might find in a zoological text-book. In the posterior end of the intestine of certain specimens are small hyolithid and brachiopod shells, and in one specimen a proboscis of another individual of *Ottoia*. Evidently this species was a carnivore—the hyolithid shells have the lid in place and so were ingested whole—and also cannibalistic.

A second, quite common species, *Selkirkia columbia* (fig. 4.23), also had a spiny proboscis, but the body was enclosed in a tube. Like *Ottoia*, it was a burrowing animal, the proboscis being extended to grip the sediment, retractor muscles pulling the tube after it, over the body. *Ottoia* and *Selkirkia*, along with three additional, rarer genera, each had a proboscis armed with spines and hooks, constructed in the same way as that of priapulids, a small phylum of burrowing marine worms living today. Most modern priapulids inhabit cold waters and anoxic muds, and in general constitute an insignificant part of benthic (bottom-dwelling) faunas. In the Cambrian they were evidently an important group of carnivorous worms, which had already undergone an evolutionary radiation, as witnessed by the five very different species found in the Burgess Shale. Apparently they were reduced in numbers and kinds later in the Palaeozoic, and their role as burrowing predators was taken over by polychaete worms; the priapulids that survive today inhabit less favourable environments.

A polychaete worm has a short head, a long trunk with many segments and a short tail region. Each segment of the trunk bore a pair of bunches of stiff bristles, or setae. An example is *Canadia spinosa* (figs 4.24, 4.25), which has the bunches of setae separated one from another by thin layers of matrix. Five additional species of these worms, each belonging to a separate genus and family, are equally well preserved in the Burgess Shale. The polychaetes are less common than the priapulids, but their diversity shows that considerable evolution had already occurred. Since the Cambrian, they have increased in numbers and kinds, and today occupy a wider range of marine habitats, burrowing, crawling, and swimming, as well as some species living in a tube. *Canadia* (see fig. 4.24) may have crawled over the sea bottom, but probably swam by means of a power stroke delivered by the outspread bunch of setae; during the recovery stroke frictional resistance was reduced probably by clumping the setae together and inclining them at a low angle to the body. *Canadia* was probably a carnivore or scavenger, but no specimens with gut contents are known. Other species were also swimmers or bottom-dwellers, and one may have lived in a burrow. None of the Burgess Shale species show jaws, groups of toothed plates comprised of tough organic material, that are called scolecodonts. Isolated jaws have been extracted with acid from Ordovician and younger rocks, and by Carboniferous times polychaetes were diverse and included many forms with jaws.

ARTHROPODA

This is the predominant group in the Burgess Shale fauna in terms of numbers of both specimens and kinds (see fig. 5.2). Arthropods are segmented animals, each segment bearing a pair of jointed limbs. The external covering, or exoskeleton, is of hardened organic material that provides both protection and an attachment-base for muscles. In some groups, for example, crabs and lobsters, the exoskeleton is strengthened by deposition of the mineral calcium carbonate. At the front of the body a number of segments are fused to form a head region, commonly bearing eyes and limbs specialized for gathering and breaking up food. The trunk consists of a series of similar segments and limbs, articulated to one another and to the head and tail. The tail includes the terminal segment, in front of which new segments are added during growth, and may in the adult consist of a number of segments fused together.

At most Cambrian localities the only remains of fossil arthropods found are the mineralized exoskeletons of trilobites. There are various kinds of these creatures in the Burgess Shale, four species being of exceptional interest because the limbs are preserved. The outstanding feature of the Burgess Shale, however, is that trilobites are far outnumbered by an extraordinary range of other species of arthropods. These are not trilobites, they did not have a mineralized exoskeleton, and are largely unique to the Shale. Because arthropods are the most numerous and varied animals on the earth today (they include insects and crustaceans), the Burgess Shale species are of immense interest as the best-known early representatives of this great group. Speculation about their relationships to each other and to younger fossil and living groups has been intense, and continues to be so. It is convenient to begin with the trilobites and then take up the other kinds.

Trilobita

In his early years, while living near Trenton Falls, New York State, Walcott found well-preserved entire exoskeletons of Ordovician trilobites. On making thin slices from these specimens, he discovered traces of the limbs. Inspired by the interest and encouragement of Agassiz, he made the nature of trilobite appendages a major theme in his research for fifty years. In his last publication on this subject, in 1921, he reviewed his earlier work with reference to the trilobites with appendages (fig. 4.26) that he had found in the Burgess Shale. The Geological Survey of Canada party was fortunate in discovering a number of excellent additional specimens of *Olenoides serratus* which gave new and better information about the limbs (figs 4.27, 4.28). The number of pairs of walking legs, their length relative to one another, the nature of the large, spiny basal joint, and the large gill structure bearing filaments, were revealed in detail. From such information I have drawn a reconstruction (fig. 4.29) which shows the animal walking. In present-day animals with many pairs of similar limbs, each pair is swung in unison, about an approximately transverse axis. Each successive pair moves in a slightly different phase of the propulsive swing backwards, or the recovery swing forwards, giving a wave of movement repeated along the series of limbs. Such a motion is inferred to be reasonable for this trilobite, and is shown in the top and side views.

The mouth, situated at the back of the plate under the head, probably faced backwards; the stomach was enclosed between this plate and the convex, central region of the head, with the alimentary canal leading back to the rear. The legs were spiny, and when a few pairs of them were curled inwards, they could grasp prey such as small worms, or scavenge carcasses. The big spiny joint at the base of each pair could have squeezed and shredded such food, and passed it forward along the mid-line of the body to the mouth. Interpreting the structures seen in the fossils in the light of the habits of similar animals today, thus leads me to suggest that *Olenoides* was a scavenger and predator, walking over, and probably digging into, the muddy bottom. It could have ploughed into soft, wet mud with the antennae swung back and the head lowered, and dug into it with the claws by alternately flexing and extending groups of legs. Movements of the limbs would move the gill branches, and keep them bathed in moving water, so that oxygen could be absorbed by the fine filaments.

The activities of a trilobite like this, walking on the mud and digging into it, would leave distinctive traces. If a layer of sand or silt was laid over the mud, these traces might be filled in and preserved on the base of the layer of sand or silt. Such traces have not been found in the Burgess Shale, because of the way it was laid down, though they must have been present in the original pre-slide environment (see fig. 3.8). Trace fossils have been described from many early Palaeozoic rocks, and some have been attributed to trilobites. In considering *Olenoides*, where so much is known about the limbs, one can suggest the kinds of traces that it may have made, and those which, though they have been attributed to trilobites, it probably did not make.

A trilobite, like an arthropod alive today, had to cast the exoskeleton at intervals during its life in order to allow the body to grow before it formed a new covering. Prior to moulting, the calcium carbonate that strengthened the exoskeleton may have been re-sorbed by the animal (as in living arthropods), so that the moulted exoskeleton may have been thin and composed mainly of organic material. The new exoskeleton formed immediately after the moulting process would have been similarly thin, before calcium carbonate was added. Thus if a newly-moulted trilobite were preserved, it might well be wrinkled and insubstantial-looking. This is the case with the specimen described by Walcott as *Nathorstia transitans*. It does not represent a distinct species, the type of a separate genus, but is a specimen of *Olenoides serratus* that shows some of the limbs as well as the exoskeleton.

Walcott found one specimen of another trilobite, *Kootenia burgessensis*, that had walking legs and gill branches preserved; no new specimens have been recovered. His collection also includes over a hundred specimens of another species, *Naraoia compacta* (fig. 4.30), that he did not think was a trilobite. The exoskeleton consists of two shields, a head and a posterior region, and limbs project from under the margins of the shields. By cutting through the shell and the thin layer of matrix below it, I was able to expose the limbs completely.

Beneath the head is one pair of antennae, and under the head and the rest of the body are a series of pairs of limbs, similar to one another, each having a walking leg and a branch bearing filaments. This series of limbs is like that of a trilobite, and I regard *Naraoia* as a trilobite even though the exoskeleton was not mineralized, and the head and tail regions were not separated by a series of segments. Recently, examination of one of the two large (27 cm in length) specimens of *Tegopelte* in Walcott's collection has shown that it also had a trilobite-like series of limbs. The exoskeleton was not mineralized, and was divided into a head region, three segments, and a large tail region. The exceptional conditions of preservation in the Burgess Shale thus show the existence of two very different kinds of trilobites with un-mineralized shells. Under ordinary conditions of preservation no trace of such soft-bodied animals would remain. It seems likely that there were many more such soft-bodied trilobites living during the Palaeozoic, only a relatively momentary glimpse of them being afforded by the Burgess Shale.

Eleven other genera, known only from the mineralized shells, occur in the Shale (see Appendix). Some specimens are complete, articulated exoskeletons (figs 1.3–6), and it is the species they represent that comprise the fauna that enabled Fritz to establish the age of the Shale. Fifteen genera at one level is a relatively large trilobite fauna for the Cambrian, reflecting both the intensive collecting and the exceptional preservation.

Arthropoda other than Trilobita

This heading embraces the largest variety of kinds in any one group of animals in the Shale (some twenty-nine genera and many more species), and the one that dominated the fauna in numbers of individuals. They were all soft-bodied, and cannot be matched among early Palaeozoic marine animals known anywhere in the world. A representative selection is illustrated and described, to show the amazing nature of this component of the Burgess Shale fossils.

Marrella splendens is the most commonly-occurring arthropod in the Shale (far more abundant than any trilobite), and a dark stain is prominent in most specimens (see fig. 3.3). In chapter 3 the various attitudes in which specimens are preserved were discussed, and their relation to the mode of burial. The exceptional specimens which are lateral (see fig. 3.4) or anterior-posterior (see fig. 3.7) compressions are most valuable in making a reconstruction. They show, for example, that the central head region was wedge-shaped, tapering downwards, but rectangular in lateral view, and how the pairs of spines curve up and out in relation to the rest of the body. Such features cannot be deduced from a dorso-ventral compression alone (see fig. 3.3), and my reconstruction (fig. 4.31) relies to a considerable extent on these unusual specimens. Because it was spiny and small, *Marrella* was thought to have floated and drifted in surface waters; however, the abundant whole specimens, engulfed in the slumps (see fig. 3.8), show that to be buried in such numbers it must have lived on the bottom. Presumably it walked on the longer front legs, the feathery second appendage being used to sweep food towards the mouth at the back of the head. The branches that bear the long, slim filaments are preserved sloping downwards, forwards or backwards. This suggests that these branches could have been swung to and fro, possibly helping the animal to swim over the bottom. Such a motion would aerate these branches, which may also have been gills.

Canadaspis perfecta is the best-known fossil species of the subclass Phyllocarida (a name referring to the leaf-like shape of part of the limb), living species of which inhabit differing environments in shallow to deep seas, and are bottom-dwellers and swimmers. The carapace (the shell that partly covers the body) is hinged on the back, the two halves, or valves, covering the anterior part of the body, so that the trunk and tail project behind it. Entire specimens, as well as empty carapaces, are abundant (in Walcott's collection half as many as the thousands of *Marrella*), and this species is the only arthropod that occurs in clusters (fig. 4.32). These clusters include complete individuals and isolated carapaces. If they were brought about by transport to the post-slide environment, it is curious that no other species of arthropod is found in such clusters. If a cluster reflects a life-habit of gathering in groups, the group having been kept together during catastrophic burial, then the presence of empty carapaces is not readily explained. This species has been intensively studied by Briggs. The part of one specimen as it was when originally split out of the rock, showed the carapace, trunk, and limbs of the left side (see fig. 4.33A). Because of the mode of burial, the rock matrix separates limbs of the left and right sides. Hence the counterpart was preserved to show these left limbs, but in the part they were dug through to reveal the limbs of the right side (see fig. 4.33B). Thus details of the appendages of each side can be obtained. Specimens buried at various angles reveal other parts of the body, which are shown in the reconstruction (fig. 4.34). *Canadaspis* was a benthic animal that dug in the mud in search of food particles, with the two valves of the carapace spread wide. The particles were sorted out, then carried under the body by the spiny limbs, the larger particles being ground up by the mandible. The leaf-like branches helped to create food-carrying currents beneath the mid-line of the body, and also absorbed oxygen.

A number of the rarer forms in the Shale have a bivalved carapace, and the appendages are at least partly known. Two of the species are phyllocarids not closely related to *Canadaspis*, whereas others are very different. *Branchiocaris pretiosa* (fig. 4.35), for example, had a great many divisions in the trunk, each of which bore a pair of thin, lobe-shaped appendages. Briggs was able to excavate the peculiar anterior antenna and second appendage, which may have been a pincer. This animal is quite unlike *Canadaspis* in appendages and numbers of segments, and cannot be called a phyllocarid.

Similar in that it had a large number of segments in the body is *Odaraia alata* (figs 4.36–38), but it may be distinguished by the large pair of eyes (borne on short stalks) and the extraordinary blades (a horizontal pair and a single vertical blade) at the posterior end. The carapace seems to have been tubular in form, so that most of the limbs were enclosed within it. By excavating specimens in stages, Briggs was able to reveal the limbs beneath the lateral lobes. These small limbs, enclosed within the carapace, could not have been used to walk on the sea bottom. Briggs postulates that the animal swam on its back, propelled by movements of the limbs, the lobed branches of which acted as a gill, so that water passed through the carapace from front to back. This current of water would have brought with it food particles, including minute floating animals, which could have been trapped and passed forward to the mouth, where the pair of mandibles could have ground them up. The tail (reminiscent of an aircraft tail) acted as a stabiliser and may have been used in steering; it may also have helped to push the animal off the bottom after resting. Though *Branchiocaris* and *Odaraia* both have a bivalved carapace and many body segments, the appendages and tail structures are quite different, thereby implying that they are not closely related. Neither can be included within the phyllocarids, nor in any other group of crustaceans (the phylum that includes the more familiar shrimps, crabs, and lobsters) living today. One other form that has a bivalved carapace, but a distinctive body and large antenna, is *Plenocaris*; it, too, is not a phyllocarid and does not fit into any group of living crustaceans.

Species of six additional genera are listed (see Appendix) as Phyllocarida, distinguished from one another by the shape and sculpture of the carapace; the body of none of them is known. As the descriptions of *Branchiocaris* and *Odaraia* have shown, animals that are not phyllocarids may have a bivalved carapace, so that the affinities of these species and genera cannot be understood until the body is known.

In Cambrian rocks minute (2 or 3 mm in maximum dimensions) bivalved shells occur, much smaller than any considered to be phyllocarids. They are regarded as having been early representatives of the abundant, living marine and fresh water ostracodes, a distinctive class of small crustaceans in which the shell encloses the body. One kind of ostracode shell is quite common in the Burgess Shale, either as separate or linked valves, but no traces of soft parts have been recognised. In Walcott's collection it is labelled as belonging to the genus *Aluta* and is given an unpublished specific name; but it has never been described.

Sidneyia inexpectans. The Burgess Shale has also yielded a variety of arthropods that have a head-shield, a series of segments in a backwardly-tapering trunk, and a tail region. The first fossil from the Shale that Walcott described was a large and striking example, *Sidneyia inexpectans*, named for his eldest son, Sidney, who found it during the first summer of quarrying (a short account of the great discovery was given by Sidney S. Walcott in 1971). A specimen 14 cm in length (fig. 4.39) shows the exoskeleton but only fragments of the limbs. Some large, jointed limbs (see fig. 4.77) occur isolated in the Shale, and Walcott regarded them as belonging to the anterior portion of *Sidneyia*. This was presumably because they were large, for he had no specimen which showed them in place. Bruton has studied the two hundred odd specimens in Walcott's collection, as well as others obtained by the Geological Survey of Canada party, and has revealed the various kinds of limbs in place beneath the exoskeleton. He has shown for the first time what the limbs were like, and that the large isolated limbs do not belong to this animal; they have now been found in place in a quite different beast, *Anomalocaris* (see below). Bruton's reconstruction of the animal is in the form of drawings and a model in plastic (fig. 4.40) made in the Museum of Palaeontology, Oslo, Norway. Such a model enables one to appreciate more clearly than in a drawing how the limbs functioned and how the animal walked, swam, and obtained food. The legs were very spiny and were well adapted for digging in the mud and grasping small prey; hyolithids and trilobites have been found in the gut contents. Thus *Sidneyia* was an active benthic predator, and may have swum, using the tail fan as well as the limbs. The head of *Sidneyia* is unusual in that it carried only the antenna and stalked eye, the other limbs being on the trunk. Bruton has shown that the structure of the trunk limb, and the way it is attached to the body, is closely comparable to that in *Limulus*, the living horseshoe crab. But *Sidneyia* did not have pincers on the front limb, as does *Limulus*, and the tail region is quite different, more like that seen in the unrelated shrimps and lobsters. Like other Burgess Shale arthropods, *Sidneyia* shows a combination of characters that exclude its being grouped with geologically younger fossils or living forms.

Emeraldella and similar genera. In contrast to *Sidneyia* are a number of distinct species in which the semicircular head-shield lacked eyes (in all but one instance), and the trunk tapered more rapidly to a terminal segment bearing a long spine. The largest of these is *Emeraldella brocki* (fig. 4.41), which had one pair of long antennae, and a series of walking legs that were spiny and had long segments (fig. 4.42). From the base of each leg branched an elongate lobe having fine lamellae around the margin, this structure being bilobed in the trunk. *Emeraldella* was another benthic animal, walking and drifting over the sea bottom and digging to find food which was then squeezed between the spiny basal joints and passed forward to the mouth.

Molaria spinifera. This animal (figs 4.43, 4.44) was commoner and smaller than *Emeraldella*, with fewer segments in the trunk and a shorter antenna. Because the animal is so small, it is difficult to reveal by mechanical means details of the limbs, but the trunk limb does not appear to have a bilobed branch. Its habits were probably like those of *Emeraldella*, but the body and spine appear to have been much more flexible. *Habelia optata* (fig. 4.45) is also small; it was quite rare, and had a tuberculate and spiny exoskeleton and a remarkable joint in the posterior spine. While studying over a hundred specimens of *Molaria*, I found seven that differed in shape and showed a large, shiny blob (presumably an eye) beside the head, and a spine the tip of which did not taper to a point but ended in a bunch of spines. These seven specimens seemed to represent a very rare but distinct form (fig. 4.46), which I named *Sarotrocercus* (brush tail) *oblita* (forgotten). Under the head was a single large, jointed limb, and under the trunk, pairs of flaps bearing flat spines. Because it had no mud filling of the gut, or walking legs (as might be expected in a benthic animal), I suggested that these specimens were carcasses of a surface floating animal (fig. 4.47). Such animals today may have large eyes.

Leanchoilia superlata. This species is set apart from the other relatively abundant arthropods by the large pair of appendages on the head, each branch of which had a long, annulated, flexible extension (fig. 4.48). The head, with its upturned snout, did not have eyes, and the short, triangular tail section bore spines. Except for a segmented and striated filling of the alimentary canal by the mineral apatite (calcium phosphate), little detail of the mid-region of the body is preserved (fig. 4.49). Lobed appendages, fringed with flat, blade-like lamellae, hang down below the body; rare specimens show a segmented inner branch, pointed at the tip, inside the front edge of the lobed branch. Bruton examined all the specimens in Walcott's and subsequent collections, including those found and described by Raymond, and constructed the model shown in fig. 4.50. The jointed inner branches of the limbs were hidden beneath the lamellate lobes, and do not seem to have been walking legs like those of *Emeraldella*. The basal portion of the limbs is not preserved, so there is no evidence of spiny processes for squeezing food and passing it forward to the mouth. Since such processes would have been of tough material, they would probably have been preserved (as in *Olenoides*, see fig. 4.27) had they existed. Thus Bruton infers that *Leanchoilia* probably swam, by flapping the lobed branches, close to the bottom, the great front appendages sensing for food. The short, jointed branches may have scraped up debris, and the mouth may have drawn in food by suction. No doubt the lamellate lobes served in swimming and as gills. As so often happens in collections of fossils, there are over a hundred that are clearly examples of *Leanchoilia*, but one that exhibits distinctive characters. This one specimen has a pair of eyes at the front margin of the head, and front appendages that are similar in construction but smaller. It is difficult to judge on a single, not very well-preserved specimen, but this appears to represent a distinct genus and species, *Actaeus armatus*, related to *Leanchoilia*. Two other specimens are not *Leanchoilia*, and though they have the eye and may have a somewhat similar frontal appendage to *Actaeus*, the limbs of the trunk show unique structures. These two are thought to represent yet another distinct genus and species, named *Alalcomenaeus cambricus*. Walcott had picked out the two rarities from his collection and labelled them as distinctive, but they were not named until Simonetta made his studies. Though *Alalcomenaeus* is a rarity in Walcott's collection, and no specimens were found by the Geological Survey of Canada party, additional specimens have been found by Collins, Briggs, and Conway Morris. Some were collected 65 m above Walcott's quarry, and *Alalcomenaeus* was found to be abundant at a locality on Mount Stephen.

Waptia fieldensis and *Yohoia tenuis*. Two animals having a rather shrimp-like tail (that is, a few hind segments without limbs and a paddle-shaped tail) are quite abundant. Of these, *Waptia* (fig. 4.51) is the most shrimp-like, having a long pair of antennae, eyes on stalks, and a small carapace, behind which the trunk projects. Some specimens show the four pairs of jointed legs beneath the carapace, and others show that on the trunk there were six pairs of gill branches, each a shaft bearing many lamellae. The six posterior segments without limbs were cylindrical, and the tail consisted of two lobes. Details of the basal parts of the limbs are not known, and thus little may be deduced about how the animal grasped and masticated food. *Waptia* was a benthic animal, perhaps perching on its anterior legs and swimming by rhythmic beating of the gills. It may have been able to dart backwards by quickly curling the back part of the body down, a mode of escape practised by living crustaceans. Presumably it fed on minute particles of organic matter on the bottom mud.

In *Yohoia*, a smaller animal (figs 4.52, 4.53) the carapace was short, the first ten segments of the trunk having curved, pointed ends, the tail a single lobe. Lobes fringed with fine hairs hung down below the trunk, and there may have been a few jointed walking legs beneath the carapace. Extraordinary is the pair of "arms" projecting in front of the carapace, which had an "elbow" joint and four "fingers" at the tip. *Yohoia* (fig. 4.54) may have acted much like *Waptia*, the anterior "arms" having been used to sense its surroundings as well as pick up food and bring it to the mouth at the front of the head.

Burgessia bella (figs 4.55, 4.56) was much more abundant than either *Waptia* or *Yohoia*, and its long posterior spine extended behind the circular carapace. Impressed into the under side of the carapace is a pattern of canals, branching from a main canal that led off the gut at the back of the head. Both gut and canal system may be preserved by being filled with sediment, and the branching system was probably for food storage or digestion. Excavation of dorsal and lateral compressions by Hughes has revealed the limbs beneath the carapace in remarkable detail. Antennae projected forward, and on the head and trunk were walking legs, those on the trunk having a lobed gill branch, those on the head bearing a long, whip-like branch (fig. 4.57). *Burgessia* probably walked on the muddy bottom in search of food, which it grasped using spiny projections on the inside of the legs; it does not appear to have been a swimmer, but may have been able to dig into the mud and partly bury itself.

At first glance the body of *Waptia, Yohoia,* or *Burgessia* might be thought to be that of a crustacean, but now that the limbs and their arrangement on the body are known in some detail, this resemblance is seen to be superficial. None was a crustacean, and each has distinctive characters which make it very different from the others.

Aysheaia pedunculata. No sooner had Walcott's description of this species, as a new kind of polychaete worm, appeared, than zoologists wrote to him pointing out similarities to the onychophorans, land animals alive today in the Southern Hemisphere. These similarities are general but striking: the onychophoran body is annulated, and has many pairs of short, stubby legs tipped with claws, and the head bears antennae. The cuticle is thin and flexible, and beneath it are layers of muscle; the limbs move by antagonistic muscles and the flow and resistance of body fluids. The living animals are found amid rotting vegetation and under logs, and can squeeze through extremely narrow passages. Their prey includes insects, and they have a powerful, horny jaw apparatus that can cut through the tough insect cuticle. Specimens of *Aysheaia* are rare, but one of them (fig. 4.58) shows how valuable it is to have both part and counterpart. By cutting through the body of the part, and the matrix below, I was able to expose the hidden left limbs and so reveal the complete series of ten on each side. The head region is best preserved in the counterpart; features of part and counterpart are combined in the drawing (fig. 4.59). Another specimen (fig. 4.60) shows the claws at the tips of the legs, the appendage on the side of the head, and, at the front of the head, a group of papillae around the mouth. A pattern of small crosses is impressed upon the right front side of the body, a pattern typical of the spicules of *Protospongia* (see fig. 4.9); other specimens of *Aysheaia* are associated with the remains of different sponges. I suggest, therefore, that the soft-skinned *Aysheaia* may have lived among clumps of sponges, which hid them from predators. The claws would have helped it grip the surface of the sponge, and the appendages would have propped the head in position while it fed on the soft parts of the sponge. It should be recognised, however, that reconstructions such as fig. 4.61 are tentative and may be proved incorrect if new evidence comes to light. *Aysheaia* was a marine animal, and far removed in time and mode of life from the modern onychophoran.

If *Aysheaia*, like the onychophorans of today, had had a jaw apparatus, it would probably have been preserved, but there is no sign of it. In this and other characters *Aysheaia* differs from onychophorans, but it is the kind of distant, marine ancestor one might expect, not only of onychophorans, but also of such important land animals as centipedes, millipedes, and insects. These latter groups, as well as the earliest onychophoran, are all known from Carboniferous rocks formed at the margins of the continents, in or near coal swamps. But fossils showing any evolutionary stages in the 230 million years between the Middle Cambrian and the Carboniferous are extremely rare. Such long gaps in time between a possible ancestor and fossils that can be assigned to a well-recognised group are frequently encountered. This particular gap and the consequent uncertainty about evolutionary relationships, raises the question of classification. Should *Aysheaia* be regarded as an arthropod? The limbs, flexible lobed outgrowths of the body, were not jointed, which means that *Aysheaia* lacked a character considered cardinal for arthropods. Yet it looks like the kind of marine ancestor from which the arthropods of the continents might have been derived; for convenience, I include *Aysheaia* among the arthropods.

CLASSIFICATION OF ARTHROPODA

The very diverse assemblage of animals, marine and terrestrial, that are included within the arthropods comprises three-quarters of all living species, and many thousands of fossil species. Spiders, crustaceans, and insects each represent a varied and distinctive group, and some scientists advocate their recognition as separate phyla, while others emphatically do not. The lively debate on arthropod classification continues in the scientific literature, but it remains difficult to fit fossils which do not obviously belong to any living group into a classification. For example, are the long-extinct trilobites, that so far as we know were a distinctive and homogeneous group of animals, to be recognised as a separate phylum? In the list of species (see Appendix), I have avoided this question by leaving the rank of the Trilobita unspecified. The Burgess Shale has yielded two trilobites, *Naraoia* and *Tegopelte*, that unlike all others so far known, do not have a mineralized exoskeleton. They have been separated informally in the list of species, but not placed in any formal category.

Among the arthropods in the Burgess Shale that are not trilobites are *Canadaspis* and *Perspicaris*, which have the limbs preserved, and may therefore confidently be classified as phyllocarid crustaceans. Whether or not other genera are placed correctly in this subclass is uncertain, because carapaces are all that is known of them. That a bivalved carapace may enclose a quite different body, which cannot be called a crustacean, is shown, for example, by *Branchiocaris* and *Odaraia*. Two other classes of Crustacea are the Ostracoda and the Cirripedia (barnacles), each having a long geological history, and one species may belong in each, with some question, as the list of species (see Appendix) indicates.

A large number of species are not placed in any phylum or class of arthropods. Some have not yet been studied in detail, but those that have differ very widely one from another. For example, *Aysheaia* stands apart because it had unbranched limbs and a flexible cuticle covering the body, not a stiff exoskeleton. *Emeraldella, Molaria, Habelia*, and, to a lesser extent, *Sarotrocercus*, show general similarities to one another, but are very different indeed from either *Marrella* or *Sidneyia* or *Leanchoilia* or *Waptia*. To place each in an extinct group of high rank, an order, or even a class, one known only from the Burgess Shale, is a rather desperate solution to classification. Størmer, after studying Walcott's specimens, thought that the non-trilobite arthropods could be brought together under one umbrella as 'trilobite-like', because each of the series of similar limbs consisted of a walking leg which had a branch fringed with lamellae, and in the head there was a single pair of antennae. Now that we know more about the varied structures of the limbs, this view cannot be maintained. What should replace it is a matter that Briggs and I have been considering. Any classification should reflect evolutionary history, but the Burgess Shale arthropods are isolated in time and space as a result of the exceptional preservation. Thus we know nothing of their ancestors and little of their descendents, so that it is difficult to discern their place in the very complex evolutionary history of the group.

ECHINODERMATA

These animals live only in the seas today, and many species (for example, sea urchins) have calcium carbonate spines projecting from the surface—hence the name, which means 'spiny-skinned'. As indicated in the first chapter, Cambrian echinoderms are different from living kinds. In the Burgess Shale only a few incomplete specimens regarded as belonging to the genus *Gogia* (see fig. 1.8) have been found. Rare examples of two small disc-shaped species called *Walcottidiscus* are also known. But most interesting are two other kinds unique to the Walcott quarry, *Eldonia ludwigi* and *Echmatocrinus brachiatus*. Several hundred specimens of *Eldonia* were found by Walcott in a single thin layer in the quarry. The fossil (fig. 4.62) has an irregular outline, and the thin film shows a central shiny coil overlaid by a pattern of branching radial lines; there is no evidence of any hard parts. Walcott thought that it was a sea-cucumber (a holothurian), a view that is maintained in the most recent study by J. Wyatt Durham. Some zoologists, however, doubt this assignment and regard *Eldonia* as a kind of jelly-fish. The central coil is considered to be the gut, and there are traces of groups of tentacles around the mouth. The radially-lined area is part of the umbrella-like body. It is thought to have floated and drifted in surface waters, and the reconstruction (fig. 4.63) shows it in such a habitat.

Echmatocrinus brachiatus (see fig. 4.64) was discovered by the Geological Survey of Canada party. It is crinoid-like, and shows what seem to be the soft parts, the tube-feet, along the arms. The plates of the arms, cup, and holdfast (originally calcite) are preserved as a thin film of pyrite, the tube-feet as a reflective layer, best seen when the specimen is immersed in liquid. The single row of plates along the arm, which appears to have borne tube-feet, suggested to Sprinkle that this species was a crinoid, but one that had an irregularly-plated cup and holdfast, rather than a stem. Crinoids are otherwise unknown before early Ordovician times, so that this much older form points to the early ancestry of the group.

HEMICHORDATA

This phylum embraces relatively small groups of living marine animals that are worm-like or form small encrusting colonies. They lack the notochord of the chordates, but some possess a gill slit or slits and resemble chordates biochemically. Fossil Graptolithina are placed here because of the resemblances in form, composition, and structure of the non-mineralized, internal and external skeleton to that of certain hemichordates. *Chaunograptus scandens* (fig. 4.65), a minute strand from which conical cups branch on alternate sides, looks like a dendroid graptolithinid, though the preservation is poor and shows no diagnostic structures. This individual appears to have grown on the outer surface of a sponge, an encrusting habit well known in dendroids. Another phylum of animals in which branching colonies are known is the Coelenterata, and since *Chaunograptus* shows no internal structures, it may equally well be a coelenterate.

Conway Morris has suggested that the worm-like species '*Ottoia*' *tenuis*, described but not illustrated by Walcott, may belong with the Enteropneusta, the worm-like hemichordates. The bases for this latter suggestion have not been explained, so that there is still doubt as to whether or not hemichordates are represented in the Burgess Shale fauna.

CHORDATA

This phylum includes all those living and fossil species that have (or are presumed to have had) a notochord. This structure is a long slim rod, composed of jelly-like material enclosed in a tough sheath, that extends usually from the base of the head along the body to the tail, affording a stiff but flexible internal support to the body.

Pikaia gracilens (fig. 4.66), originally described by Walcott as a worm, is represented by sixty specimens. Conway Morris regards the longitudinal strip along the upper edge of the body as the notochord. The median region of the body is reflective and probably incorporates the gut. Behind the head are zig-zag segmental boundaries reminiscent of the characteristic musculature of fish. Conway Morris's interpretation implies that *Pikaia* is the earliest chordate known, and these specimens are unique among all the fossils considered to be chordates in that a trace of the notochord is preserved. Vertebrates (fish, amphibians, reptiles, and mammals) constitute the vast majority of the chordates, and have a skeletal system, including the vertebral column, composed of cartilage or bone. Bits of fossil bone have recently been found in latest Cambrian or earliest Ordovician rocks and are evidence of the existence of vertebrates, so that *Pikaia* is important in showing the early origin of chordates.

MISCELLANEOUS ANIMALS

This category includes an assortment of species, each of which has a combination of structures unknown in any fossil or living species, and so cannot be placed in any recognised phylum. The number of genera in this assortment is over one-third that of the arthropods (see fig. 5.2), and the numbers of individuals known in most cases is few, the exception being the one hundred and forty specimens of *Wiwaxia*. Some of these animals (or parts of them) were described by Walcott, particularly in his work on worms, and one or two were segregated in his collection but not commented on. The new investigation has revealed the strangeness of some of these forms, described below; others remain to be studied.

The extraordinary *Opabinia regalis* (figs 4.67, 68) has attracted the attention of various investigators, and speculations have been advanced on its nature and its possible relationships within the arthropods. A detailed examination of the fossils, however, has shown that its relationships are not as supposed. A new specimen (fig. 4.69) collected by the Canadian Geological Survey party demonstrates quite clearly that it does not have pairs of jointed legs and so cannot be an arthropod. It also shows how the frontal process could be curved back to bring food, held in the spines at the tip, to the backwards-facing mouth. Along each side of the body was a row of overlapping lobes, on the outside of which were lamellate structures, presumably gills. These structures appear as narrow strips running longitudinally, and may be seen to alternate with the lobes in the specimen shown in fig. 4.68. There was a group of five eyes on the head, and at the end of the body three pairs of upwardly directed blades. The body was segmented and flexible, and the creature could have swum over the bottom mud by rhythmic movements of the side lobes, while the frontal process explored for organic debris as food (see fig. 4.70). The blades would have assisted in steering, and their movement and that of the side lobes would have kept water circulating between the lamellae of the gills. *Opabinia* may have descended from the kinds of segmented animals that also gave rise to arthropods and certain kinds of worms.

A new discovery is *Dinomischus* (fig. 4.71). Only one specimen was collected by Walcott, one by Raymond, and one by the Royal Ontario Museum party, so that it is extremely rare. The cup-shaped body was held in place by the stem, which had the tip swollen to form a holdfast, embedded in the mud. Curious arm-like structures on the cup may have borne fine hair-like processes which lashed to and fro in groups to create currents of water. Thus minute food particles in suspension could have been channelled to the mouth and then digested. One may imagine these animals scattered thinly about, attached to the sea floor, swaying in the bottom currents that brought them food and oxygen. The smooth cup and flattened arms did not have calcareous plates supporting them, like those of animals of similar habit, such as the eocrinoids (see fig. 1.8). The relationships of this species are obscure, but reflective traces of the soft parts in the cup suggested to Conway Morris some analogies with living entoprocts, a minor group of tiny marine and fresh water animals.

It would have been difficult to imagine an animal as bizarre as *Hallucigenia* (figs 4.72, 4.73). Preparation of some of the forty specimens (included by Walcott in a supposed species of worm) showed the elongate body, on the lower side of which were seven pairs of sharp-pointed spines. On the upper side of the body were seven tentacles, each bearing at the tip a snapper-like structure. The rear end of the body curved upwards and bore a group of smaller tentacles; the front end is preserved only obscurely, the outline vague, presumably as a result of decay. *Hallucigenia* may have moved over the bottom mud on its stilt-like pairs of spines, grasping food with its tentacles; but how the food reached the alimentary canal that traversed the body is uncertain. A unique specimen in Raymond's collection shows a group of individuals associated with a mat of organic material, suggesting that they were scavenging a corpse. The puzzling nature and possible relationships of this species need no emphasis!

Individual ribbed scales and long spines of *Wiwaxia corrugata* occur on many slabs of shale collected from Walcott's quarry. Entire specimens (fig. 4.74) show that the differently-shaped scales were arranged in overlapping symmetrical rows along the sides and top of the body, and that a row of spines projected up along each side. The under side of the body may have been a flexible skin lacking scales, enabling the animal to crawl along the bottom. Rare specimens show a feeding apparatus of two rows of backwardly-directed teeth, which indicate that *Wiwaxia* may have been able to scrape up food, or scavenge. The upwardly-projecting spines may have been to protect the animal from attack, as suggested by the occasional specimen in which the spines are broken. Many snails have rows of tiny teeth which can be pulled to and fro, allowing them, for example, to scrape algae from rock surfaces. This similarity is not enough to suggest that *Wiwaxia* might be a strange kind of mollusc, and Walcott, because of its scales, described it as a worm. Dissolution in acid of Lower Cambrian limestones from localities in various parts of the world has produced residues that include phosphatic spines and scales like those of *Wiwaxia*. It may be, therefore, that *Wiwaxia*-like animals were widespread during the Cambrian; the hollow structure of these isolated scales is unlike that known in other kinds of animals.

In one of his early papers Walcott described, as *Peytoia nathorsti*, a circular, plated structure which was gently convex (fig. 4.75). The plates are divided into four equal sections by wider plates, any two wider plates being separated by seven narrower plates; all the plates have blunt teeth at the inner end which project into a central opening. Because among living animals the jelly-fish are circular in shape and have a four-rayed symmetry, the fossil *Peytoia* has been regarded as a jelly-fish. In other articles in which he described arthropods, Walcott included two isolated limbs. One, which had been named *Anomalo-*

caris canadensis (fig. 4.76), had a pair of spines on the inside of each joint. A limb of this kind had been found in other places in western Canada before Walcott reported it from the Burgess Shale. *Anomalocaris* was thought to be not the limb of an animal, but more likely the back part of the body of a *Canadaspis*-like form (see fig. 4.33), which had become isolated from the bivalved shell. The second type of limb (fig. 4.77) Walcott found in the Burgess Shale was distinguished by having a long blade on the inner side of each joint, each blade bearing spines and ending in a pointed, curved tip. No name was given to these specimens, but Briggs referred to them as 'appendage F'. Walcott thought these big limbs must have belonged to *Sidneyia* (see fig. 4.39), the largest arthropod he had found. However, Bruton's re-study of all the specimens of *Sidneyia* showed that limbs of this type are not attached to it. Thus two kinds of isolated limbs, preserved because the outer covering was thickened, occur in the Shale, the body of the animal to which either belonged being unknown.

Only recently has the problem presented by these limbs been solved, as a result of investigations by Briggs and myself of some rare specimens in the Walcott collection and one found by the Canadian Geological Survey party. The latter (fig. 4.78) has a pair of limbs of the *Anomalocaris canadensis* type, symmetrically arranged and attached to the narrower front part of the body. Behind this the body widens and then narrows back, a series of lobes along each side. The few specimens in Walcott's collection are better preserved and similar in form (figs 4.79, 4.80), but to our surprise have a pair of limbs of the 'appendage F' type attached at the front end. Even more surprising was the discovery, when trying to expose the limbs at the front end (fig. 4.79), of the circlet of plates called *Peytoia*. Because of their position, these plates must be the mouth parts of the animal, the armature of teeth around the central opening being for grasping and breaking up food caught and held by the limbs. Limbs and mouth-parts had a tough, hardened (but not mineralized) outer covering, and so were preserved after the softer body-covering decayed; complete specimens are preserved only in the special conditions of the Shale.

The overlapping row of triangular lobes along each side of the body was thin and flexible, and could probably have been moved up and down rhythmically to form a wave travelling back in unison along each side, thereby enabling the animal to swim over the bottom in search of food (fig. 4.81C). Lamellar gill structures were attached along the body above the lobes and beneath the dorsal cuticle. The dimensions of large isolated specimens of the limbs, and of *Peytoia* circlets of plates, show that the whole animal may have been almost half a metre in length. It is the largest creature known from the Burgess Shale, and was bigger than any other animal known from Cambrian rocks anywhere. It must have been a formidable predator, swimming over the bottom with the front limbs ready to catch prey (fig. 4.81). The spiny limbs and toothed jaw would have combined to tear up the victim. No gut contents are known, and while *Anomalocaris* was big enough at least to wound trilobites, it may well have preyed mainly on the far more abundant animals without mineralized skeletons. Although this limb was jointed, there was only one pair of them, and the animal can therefore hardly be called an arthropod. The body was segmented, the jaw apparatus unique, and the manner of swimming recalls the way in which the squid uses the long lateral fin. Such a strange combination of characters means that this beast cannot be placed in any recognised phylum.

When he described the circlet of plates in 1911, Walcott gave it a new generic name, *Peytoia*. The isolated limb had been called *Anomalocaris* in 1892. A third generic name, *Laggania*, was given by Walcott in 1911 to a specimen which can now be recognised as an incomplete specimen of the whole animal. So which generic name should be used? International rules adopted to deal with this kind of problem specify simply that the oldest name be used, in this case *Anomalocaris*, despite its having been applied originally to only a part of the animal. This kind of difficulty over names is familiar to palaeobotanists, in cases where isolated leaves, fruits, or wood have been described and named before a more complete specimen has been found. The problem of *Anomalocaris* illustrates how an apparently complete specimen, *Peytoia*, may prove to be only a portion of an animal, and how misleading speculation on affinities may be. Jelly-fish have long been thought to be a characteristic element in the Burgess Shale fauna, but the basis for this view has now gone.

Amiskwia sagittiformis (fig. 4.82), known from only five specimens, had any elongate body with an anterior pair of tentacles, a lateral pair of fins, and a single posterior fin. Reflective patches and strips show details of the soft parts, including the intestine, nerve chord, and ganglion in the head; other reflective lines may represent blood vessels. In his re-description of this species, Conway Morris concluded that though it is worm-like, it is not an arrow-worm (Phylum Chaetognatha, a small group of marine, and generally planktonic, animals) nor a ribbon-worm (Phylum Nemertea), certain species of which inhabit deep ocean waters. The systematic position of *Amiskwia* is thus uncertain; Conway Morris postulates that it probably swam actively in the surface waters.

Only a single, tiny specimen of *Nectocaris pteryx* (fig. 4.83) is known. Walcott noted and photographed it, but the first description is by Conway Morris. The body was laterally compressed, and the head, which was partly enclosed in a pair of oval shields, bore appendages and a pair of large eyes (preserved as a reflective patch). The elongate trunk had a fin, supported by rays, along the dorsal and ventral sides. The long, presumably muscular, trunk with its fins suggests to Conway Morris a swimming animal, a predator having large eyes and anterior appendages to aid it in grasping prey.

Fig. 4.1. *Morania confluens*, one of the branching strips of irregular, perforated sheets that Walcott described.

Fig. 4.2. *Marpolia spissa*, a portion of a slab covered with broken fragments.

Fig. 4.3. *Wahpia virgata*, the only specimen of this branching alga that Walcott described.

(A)

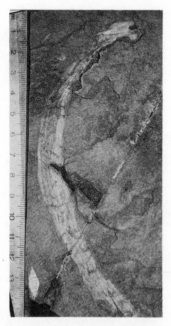

Fig. 4.5. *Leptomitus lineata*, an elongate, thin-walled sponge that had long, thin spicules.

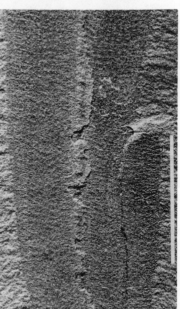

(B)

Fig. 4.4. *Vauxia gracilenta*, an example of this branching sponge described by Walcott, A, and an enlargement showing the pattern of the network in the wall, B.

Fig. 4.6. *Pirania muricata*, showing the hollow central stem made of close-packed, bolt-shaped spicules, with bunches of long spicules attached to it.

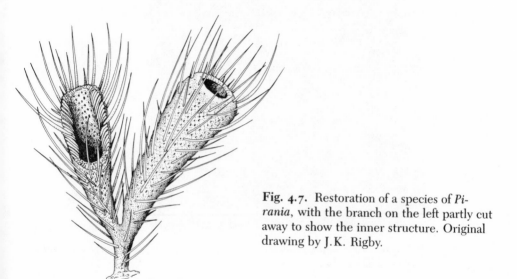

Fig. 4.7. Restoration of a species of *Pirania*, with the branch on the left partly cut away to show the inner structure. Original drawing by J.K. Rigby.

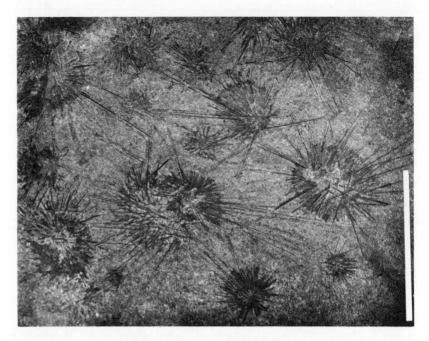

Fig. 4.8. *Choia ridleyi*, a sponge with radially-arranged long spicules. Photograph by J. K. Rigby.

Fig. 4.9. *Protospongia hicksi*, part of the wall thereof, formed by spicules arranged in a series of squares.

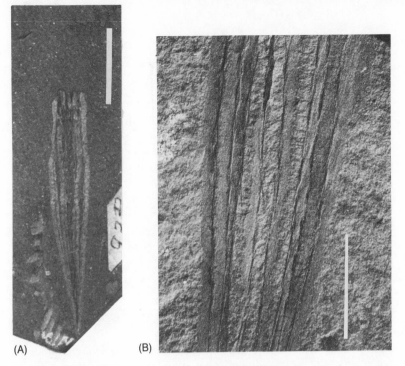

(A) (B)

Fig. 4.10. *Takakkawia lineata*, one of Walcott's entire specimens, A; enlargement showing the spirally twisted lines of spicules, B (scale bar 0.5 cm). Photographs by J.K. Rigby.

Fig. 4.11. *Micromitra burgessensis*, the shells attached, as they were in life, to the sponge *Pirania*.

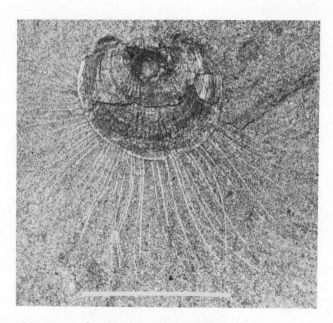

Fig. 4.12. *Micromitra burgessensis,* an isolated example showing the long hair-like processes of the mantle projecting from between the valves of the shell.

Fig. 4.13. *Lingulella waptaensis,* an elongate brachiopod shell showing concentric lines of growth.

Fig. 4.14. *Diraphora bellicostata*, the two valves of the strongly-ribbed shells linked together.

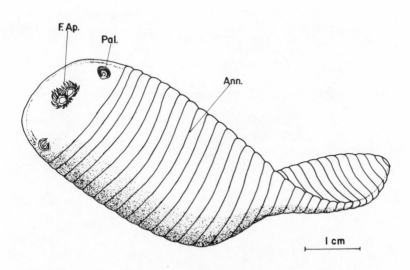

Fig. 4.15. A reconstruction of *Odontogriphus omalus*, the anterior portion showing the feeding apparatus (F. Ap.) and sensory lateral palps (Pal.) on the under side, the posterior portion of the annulated body (Ann.) twisted to show the upper side. From Conway Morris, 1976, *Palaeontology*, vol. 19, p. 206, text-fig. 3.

(A)   (B)

Fig. 4.16. *Mackenzia costalis*, taken at low angle, A, and with the radiation reflected, B. Both photographs show the elongate body and ridges that are thought to be made by the ends of radial partitions.

Fig. 4.17. *Scenella* sp. ind., low, conical shells referred to this genus of molluscs, which were common in the Burgess Shale and widespread in Cambrian rocks.

4.19   4.18

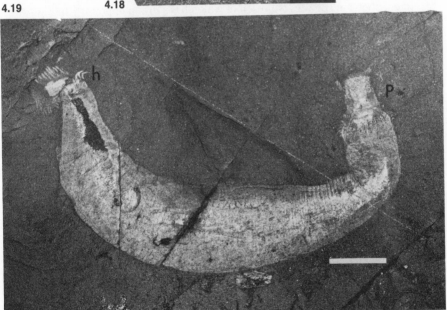

Fig. 4.18. *Hyolithes carinatus*, a complete, and therefore unusual, speci-men of the conical shell and the lid. The curved blade-like strip on each side helped to prop the animal in position on the sea floor.

Fig. 4.19. *Ottoia prolifica*, an entire, well-preserved specimen, showing the proboscis (p), annulations on the body, and the posterior hooks (h). Pho-tograph by Conway Morris, specimen collected by Geological Survey of Canada party.

Fig. 4.20. *Ottoia prolifica*, an enlargement of the proboscis of another specimen, showing spinules (sp, each of which has a group of points), spines (s), and hooks (h). Photograph by Conway Morris, specimen collected by Geological Survey of Canada party.

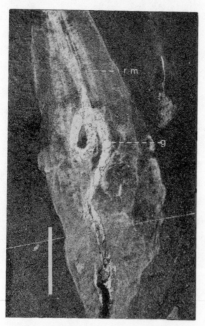

Fig. 4.21. *Ottoia prolifica*, part of a specimen showing the reflective gut (g) down the centre of the body, coiled in a loop, and the less strongly reflective strips beside it that have been interpreted as retractor muscles (rm). Photograph by Conway Morris.

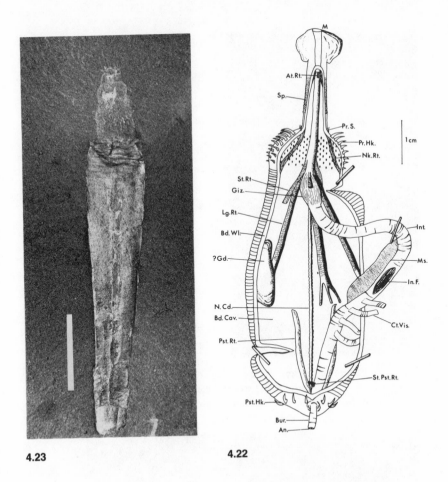

4.23

4.22

Fig. 4.22. *Ottoia prolifica*, hypothetical dissection, cut along the dorsal mid-line and the body wall (Bd. Wl.) and pinned back to show the internal organs in the body cavity (Bd. Cav.). The proboscis, which is fully everted (more so than in fig. 4.20), has the mouth (M) at the tip, and shows spinules (Sp.), proboscis spines (Pr. S.), and hooks (Pr. Hk.). Muscles that retracted the proboscis were anterior (At. Rt.) and at the 'neck' (Nk. Rt.). The gut includes the muscular gizzard (Giz.), and the intestine (Int.) with internal furrows (In. F.) in the wall, and ends in the eversible bursa (Bur.). The gut was suspended in the body cavity by fibrous mesenteries (Ms.) and cutaneovisceral muscles (Ct. Vs.). The presumed urinogenital system (Gd.), various retractor muscles (longitudinal, Lg. Rt.; posterior, Pst. Rt.; and short posterior, St. Pst. Rt.), and the dorsal nerve chord (N. Cd.) are also shown. From Conway Morris, 1977, *Fossil priapulid worms*, Special Papers in Palaeontology, No. 20, p. 12, text-fig. 6.

Fig. 4.23. *Selkirkia columbia*, an unusual specimen showing the soft parts (most specimens are empty tubes, as in fig. 4.64). The proboscis is not completely extended, and has spines at the tip and sides; the tube is crumpled just behind it, the remainder showing a trace of the gut. Photograph by Conway Morris, specimen collected by Geological Survey of Canada party.

Fig. 4.24. *Canadia spinosa*, complete specimen with a pair of tentacles (t) at the anterior end, and a reflective median strip, which is the trunk from which lateral lobes project. The bunches of setae are attached to these lobes. Photograph by Conway Morris.

Fig. 4.25. *Canadia spinosa*, part of a specimen showing one bunch of the notosetae (no) that covered the dorsal side of the body, and a series of bunches of neurosetae (nu) that projected from the side of the body and were used in walking and swimming. Scale bar 0.5 cm. Photograph by Conway Morris.

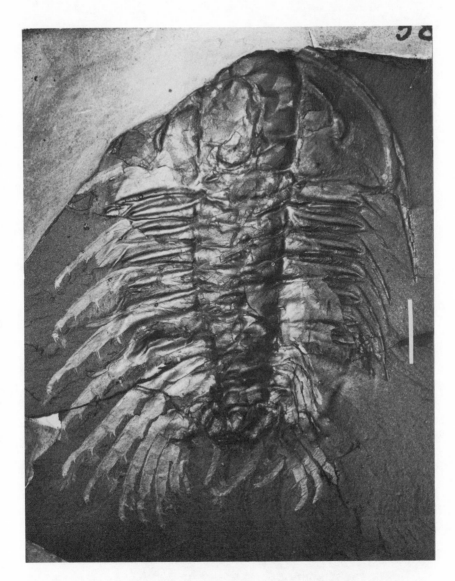

Fig. 4.26. *Olenoides serratus.* Much of the mineralized exoskeleton has been stripped off, so that a mould of the under surface, and the limbs, are visible. The limbs are displaced back and to the left of their original position in relation to the exoskeleton. The limbs on the right are bent to point backwards. The radiation was directed steeply down onto the specimen, which was tilted slightly, resulting in the maximum reflection from the limbs.

Fig. 4.27. *Olenoides serratus*, a specimen split through the limbs and the under side of the exoskeleton, photographed under water to maximise reflection. The limbs are displaced to the right relative to the exoskeleton. Beneath the curving edge of the tail, the spiny inner segments (i) of the legs are preserved.

Fig. 4.28. *Olenoides serratus*. The filaments (f) of the gill branch (g) of the limb at the top have been exposed to show how the long, slim filaments extend back below the next two legs. The radiation was directed at a low angle across the specimen from the left side, to emphasise the changes in level between bristles of the leg and between gill filaments.

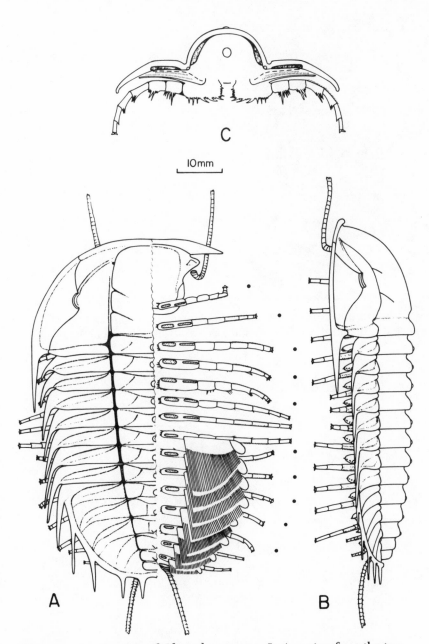

Fig. 4.29. A restoration of *Olenoides serratus*. In A, a view from the top,
the right half of the exoskeleton has been cut away to show the plate beneath the head and the limbs, and the gill branches of the first seven limbs
have been removed to show the legs below in attitudes of walking. The dots
are opposite the tips of the limbs that are on the sea bottom. B is a side
view showing the attitudes of the legs when walking. C is a cross-section of
the body showing one pair of limbs in posterior view; the full length of the
antennae and posterior cerci is not shown. From Whittington, 1975, *Fossils
and Strata*, vol. 4, pp. 124–25, figs 25, 26.

Fig. 4.30. *Naraoia compacta.* Crushed slightly obliquely, the antenna on each side of the head is seen, and limbs along the left side and to the rear.

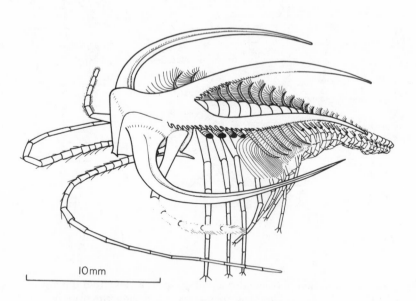

10mm

Fig. 4.31. Restoration of *Marrella splendens.* Only the walking legs of the near side are shown, and many of the front and back gill branches have been cut off so that the leg branches may be seen. Compare figs 3.3 and 3.7 for specimens preserved in various attitudes, on which this drawing is based. From Whittington, 1971, *Geological Survey of Canada Bulletin* 209, text-fig. 5.

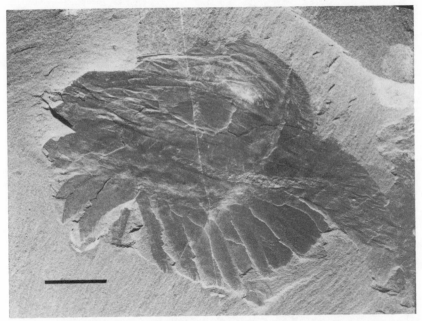

(A)

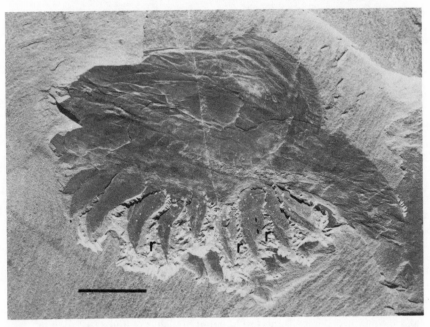

(B)

Fig. 4.33. *Canadaspis perfecta.* A shows the part of a complete specimen as it was split out of the rock, and reveals the limbs of the left side. B shows the same specimen after cutting down into the rock to reveal the buried limbs of the right side (r). The limbs of the left side are preserved in the counterpart (not shown). Photographs by D.E.G. Briggs.

Fig. 4.32. *Canadaspis perfecta*, showing a cluster of small carapaces, with one specimen of *Naraoia* (N), on a slab of rock. Photograph by D.E.G. Briggs.

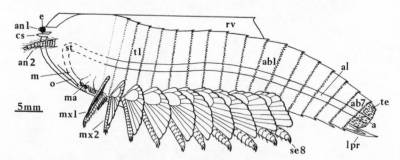

Fig. 4.34. A restoration of *Canadaspis perfecta*. The left valve of the carapace is removed, the right (rv) outlined. The left appendages shown are the first (an 1) and second (an 2) antennae, the mandible (ma), the first (mx 1) and second (mx 2) maxilla, and segmented branches of limbs (se). Behind the head region, which has a spine (cs), the eye (e), a lip (o), a mouth (m), and a stomach (st), the body is divided into a thorax of eight segments (t), and an abdomen of seven segments (ab). The body is traversed by the alimentary canal (al) ending at the anus (a) in the terminal portion (te) of the body. The segment in front of this portion has a projection (1pr) on the under side. From Briggs, 1978, *Philosophical Transactions of the Royal Society, London*, vol. B 281, p. 445, fig. 27.

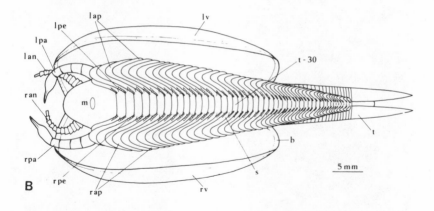

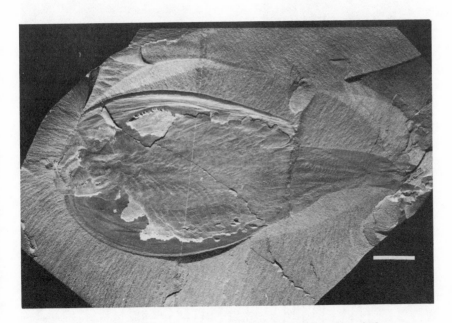

Fig. 4.35. *Branchiocaris pretiosa*. The photograph, taken from below, shows the right antenna (ran, in the reconstruction above) and principal appendage (rpa) at the front, and the many-segmented body and lamellate appendages (rap, lap) that are exposed beneath the left and right valves (lv, rv) of the carapace. The reconstruction, drawn from this and other specimens, also shows the basal portion (pe) of the appendages, the mouth (m), trunk segment 30 (t-30), the terminal portion of the body (t), the border of carapace (b), and the striations on the basal portions of the appendages (s). From Briggs, 1976, *Geological Survey of Canada Bulletin* 264, p. 7, text-fig. 2B.

Fig. 4.36. *Odaraia alata*, a view from above, showing the large eye (e), the carapace (c), and the many segments of the body, each bearing a lobed branch, or gill (g). Photograph by D.E.G. Briggs.

Fig. 4.37. *Odaraia alata*, posterior portion of the carapace (c) of another specimen, and the body projecting behind it, ending in the tail with its horizontal (h) and vertical (v) blades. Photograph by D.E.G. Briggs.

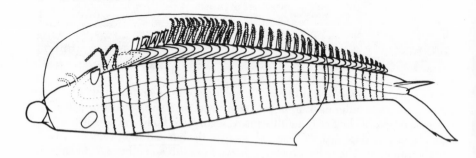

Fig. 4.38. *Odaraia alata*, restoration in right lateral view, as it may have looked when swimming on its back; only the outline of the carapace is shown. From Briggs, 1981, *Philosophical Transactions of the Royal Society, London*, vol. B 291, fig. 103d.

Fig. 4.39. *Sidneyia inexpectans*. Fragments of the appendages emerge on each side of the head. The gut, filled with undigested fragments, runs down the middle of the narrow back part of the body. Photograph by D.L. Bruton.

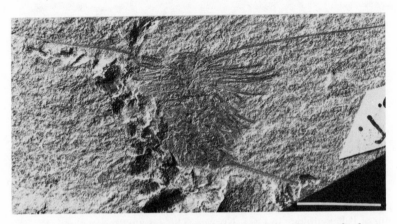

Fig. 4.41. *Emeraldella brocki*, from the under side, showing the long antenna, the many long walking legs, and the tail spine.

Fig. 4.40. Three views of the model of *Sidneyia inexpectans* made at the Museum of Palaeontology in Oslo, Norway. The antenna and eye emerge from the edge of the head. The upper and middle views show the manner of walking. In the lower (front) view the plates of the gills are visible between the legs of the posterior portion of the body. Gill branches and legs, when swinging to and fro, may have enabled the animal to swim, and at the same time currents of water would have bathed the gills.

Fig. 4.42. *Emeraldella brocki*, the head and three segments of the body, showing spines on the legs and a lobe with fine marginal lamellae on the left of the first segment.

Fig. 4.43. *Molaria spinifera*, top view of a typical specimen.

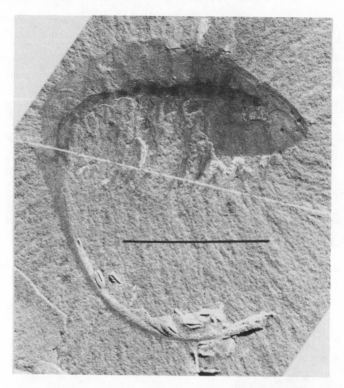

Fig. 4.44. *Molaria spinifera*, flattened from the side, showing how the body and spine could be curled. The dark strip is the alimentary canal, and below the body are the limbs.

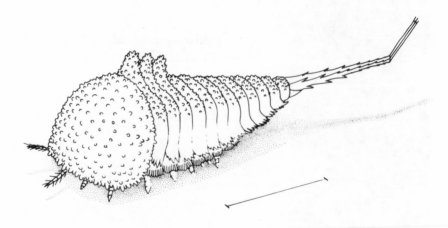

Fig. 4.45. A restoration of *Habelia optata*, an animal rather like *Molaria spinifera* but with tubercles and spines on the head and trunk, and a joint in the posterior spine. From Whittington, 1981, *Philosophical Transactions of the Royal Society, London*, vol. B 292, fig. 130.

Fig. 4.46. *Sarotrocercus oblita,* one of the rare specimens, showing the eye (e) at the side of the head, and spines (s) at the tip of the tail spine. Scale bar 0.5 cm.

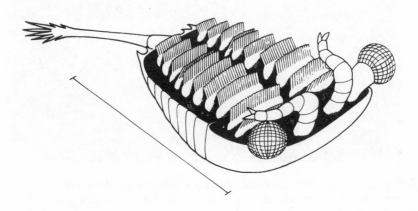

Fig. 4.47. *Sarotrocercus oblita,* a reconstruction portraying the animal as it might have floated in surface waters. From Whittington, 1981, *Philosophical Transactions of the Royal Society, London,* vol. B 292, fig. 131.

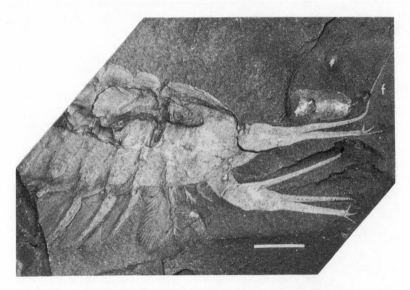

Fig. 4.48. *Leanchoilia superlata*, viewed obliquely from below, showing the head and first few segments of the body, with the lobed appendages hanging down below. A pair of large branched appendages project forward from the head, and the flexible extension (f) of one branch is curved upwards.

Fig. 4.49. *Leanchoilia superlata*, the entire animal viewed from the side, with the great appendages (g) folded backwards and downwards, the flexible tips (t) curving down below the lobed appendages of the body in what may have been a 'feathered' position of the great appendages when swimming. Dark blobs along the body behind the head are the mineral apatite in the gut.

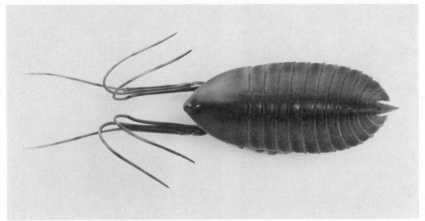

Fig. 4.50. *Leanchoilia superlata*, three views of the model made at the Museum of Palaeontology in Oslo, Norway. The underneath and side views show how the lobed appendages could have been moved to and fro in swimming.

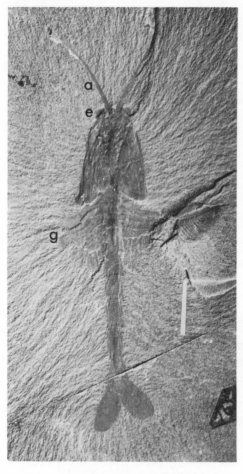

Fig. 4.51. *Waptia fieldensis.*
The antennae (a) and eye lobe
(e) are at the front of the carapace, and behind it are the gill
branches (g), the posterior segments, and the bi-lobed tail.
Part of a *Marrella* is beside the
gill branches on the right. Photograph by C. P. Hughes.

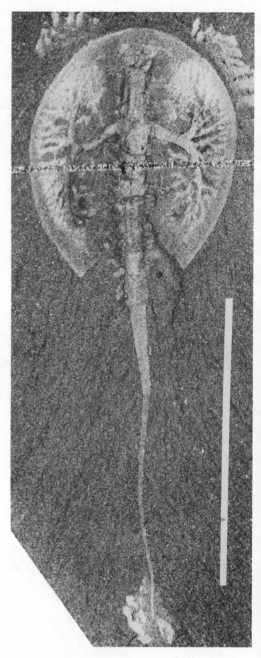

Fig. 4.55. *Burgessia bella.* Beneath the circular carapace, in
this photograph by C. P.
Hughes, taken in reflected
light, is the pattern of canals
branching from the gut, and the
long, flexible posterior spine.

4.53

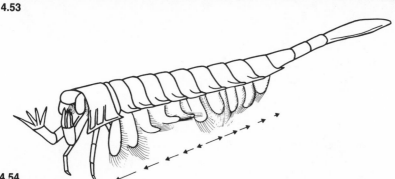

4.54

Fig. 4.52. *Yohoia tenuis*, showing the curved ends of the trunk segments (t) and the tail. Parts of limbs are visible below the head (h) and trunk.

Fig. 4.53. *Yohoia tenuis*, showing the large 'arms' (a) in front of the head, flexed at an 'elbow' joint (e), and having four spines at the tip. Traces of limbs are seen below the body, and the tail lobe is broken.

Fig. 4.54. An impression of *Yohoia tenuis* as it may have looked when swimming by means of movements of the lobes beneath the body (the arrows show the directions of the rhythmic movements). How many pairs of jointed legs were present is uncertain because the limbs are poorly preserved in these tiny specimens. From Whittington, 1974, *Geological Survey of Canada Bulletin* 231, text-fig. 5.

Fig. 4.56. *Burgessia bella.* The carapace is bent down in this specimen and distorted by flattening. The antennae and pairs of jointed legs are well preserved, the legs are displaced backwards relative to the carapace, and the posterior spine is broken. Photograph by C.P. Hughes.

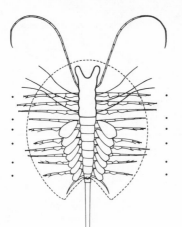

Fig. 4.57. A restoration of *Burgessia bella* in top view, showing the limbs in walking position. The dots are opposite the tips of the legs which are on the sea bottom. The outline of the carapace is dashed. From C.P. Hughes, 1975, *Fossils and Strata*, vol. 4, p. 431, fig. 6.

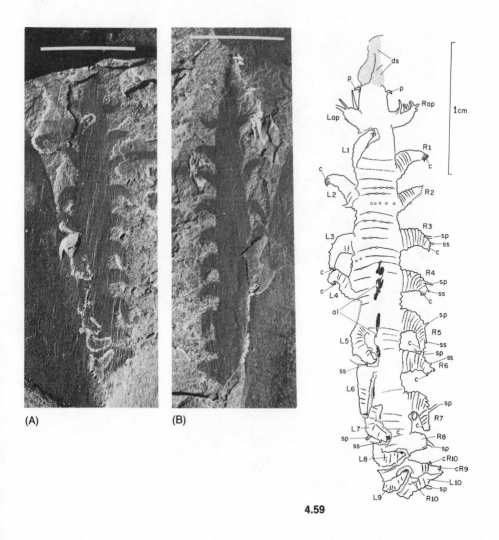

(A)   (B)

4.59

Fig. 4.58. *Aysheaia pedunculata*, the part, A, and counterpart, B, of the specimen originally described by Walcott. In the part several of the left limbs were bent under the body, and have been exposed. In the counterpart the head end with its branched appendage is better preserved than in the part, but the tip is obscured by a dark stain.

Fig. 4.59. *Aysheaia pedunculata*, a drawing combining the features seen in the part and counterpart of fig. 4.58. The R and L prefixes indicate right or left appendage (ap), or the limbs, numbered 1 to 10. Large (sp) and small (ss) spines occur on the limbs, which have claws (c) at the tip. In front of the head is a dark stain (ds), and obscure papillae (p) are visible; parts of the alimentary canal (al) are shown by reflected light as darker patches. From Whittington, 1978, *Philosophical Transactions of the Royal Society, London*, vol. B 284, p. 174, fig. 4.

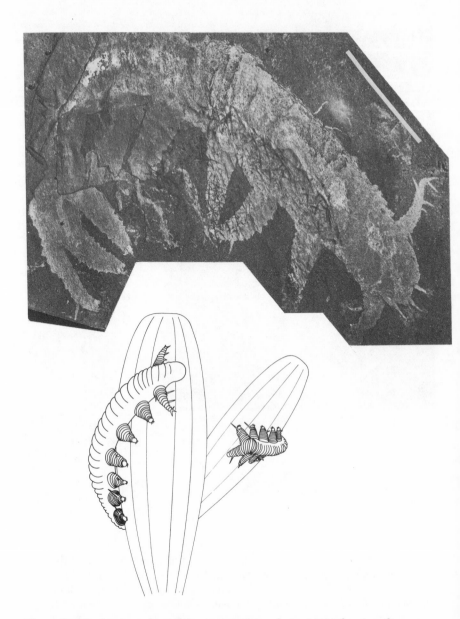

Fig. 4.60. *Aysheaia pedunculata*, an exceptional specimen showing the papillae around the mouth, the branched appendage on the side of the head, a reflective trace of the alimentary canal, and the groups of claws at the tips of the back limbs. The black lines in a pattern of crosses on the right side are sponge spicules.

Fig. 4.61. *Aysheaia pedunculata*, as it may have appeared when clinging to a branching sponge, the claws helping it to adhere. The appendages on the side of the head are propping it in position, and the papillae may have wounded the sponge so that the soft parts could be sucked out. From Whittington, 1978, *Philosophical Transactions of the Royal Society, London*, vol. B 284, p. 193, fig. 90.

Fig. 4.62. *Eldonia ludwigi.* The shiny loop is the gut, and around the mouth are two bunches of tentacles (t). Radial lines run from the centre, inside the loop, out to the irregular margin.

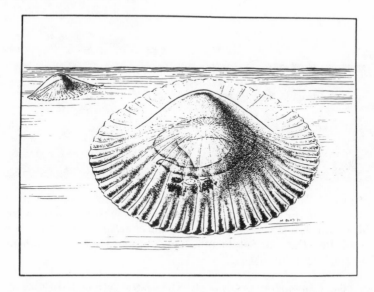

Fig. 4.63. *Eldonia ludwigi*, reconstruction by J. Wyatt Durham, showing the animal as a surface-floating holothurian. The body is bell-shaped, the radial lines are thin sheets by which the gut, inside the body cavity, is held in place. The mouth, with tentacles, and the anus opened on the under side. From *Journal of Paleontology*, 1974, vol. 48, p. 754, text-fig. 2.

4.64

s

s

s

4.66   **4.65**

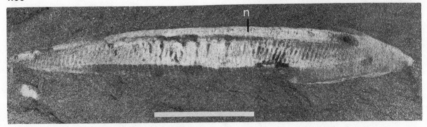

n

Fig. 4.64. *Echmatocrinus brachiatus*, a unique specimen collected by the Geological Survey of Canada party. The conical cup passes down into the holdfast, which is attached to an empty tube of *Selkirkia*. The arms extend upwards from the cup, and the tube-feet curve out sideways from them. Photograph by J. Sprinkle.

Fig. 4.65. *Chaunograptus scandens*. The central portion of the specimen is probably of the sponge *Leptomitus* (see fig. 4.5). On each side is a thin strand (s) from which small cups branch alternately; these are thought to be the remains of a graptolithinid.

Fig. 4.66. *Pikaia gracilens*, showing the notochord (n), below which is a central reflective strip, probably incorporating the gut. In the trunk (mid and right-hand parts of the body) are zig-zag boundaries between segments, as in the characteristic musculature of fish. Photograph by Conway Morris.

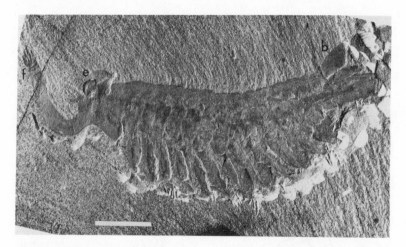

Fig. 4.67. *Opabinia regalis*, a laterally-flattened specimen showing the eyes (e) and frontal process (f) on the head and two of the upstanding blades (b) at the end of the body.

Fig. 4.68. *Opabinia regalis*, a specimen flattened vertically, viewed from underneath. Behind the dark patches that are the eyes (e), the lateral lobes and gills alternate, projecting out from the body. The dark middle line is a trace of the gut.

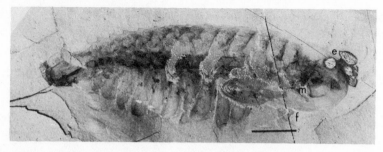

Fig. 4.69. *Opabinia regalis*, photographed immersed in alcohol, showing the group of eyes (e) and the frontal process (f) curved back. Beside the base of the process, outlined by a darker line, is the backwards-facing mouth opening (m). The dark strip down the centre of the body is the alimentary canal. There are no jointed limbs below the lateral lobes. Specimen collected by Geological Survey of Canada party.

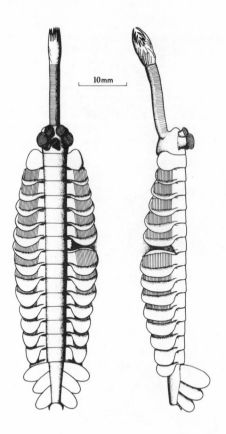

10mm

Fig. 4.70. *Opabinia regalis*, top and side views of a reconstruction. From Whittington, 1975, *Philosophical Transactions of the Royal Society, London*, vol. B 271, p. 34, fig. 82.

Fig. 4.71. *Dinomischus isolatus*, showing the cup, the arm-like structures above it, and the upper portion of the stem. Scale bar 0.5 cm. Photograph by Conway Morris.

Fig. 4.72. *Hallucigenia sparsa*, photographed in reflected radiation, show-ing the spines (s), some of the tentacles (t), the short tentacles (st), and the trunk curving up at the back. Photograph by Conway Morris.

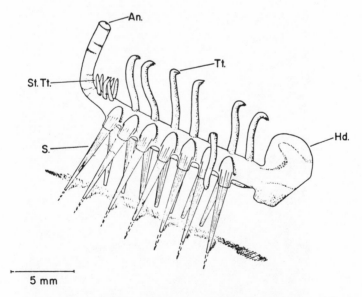

Fig. 4.73. *Hallucigenia sparsa*, reconstruction by Conway Morris. The head (Hd.) is only vaguely known, and the body stands on seven pairs of spines (s.) and has shorter (St.Tt.) and longer (Tt.) tentacles on the upper side, and the anal opening (An.) at the tip. From *Palaeontology*, 1977, vol. 20, p. 628, text-fig. 2.

Fig. 4.74. *Wiwaxia corrugata*, an almost symmetrical specimen. The long spines on each side are not all visible, and the scales are displaced so that only a few from the various paired rows are recognisable. Photograph by Conway Morris.

Fig. 4.75. *Anomalocaris nathorsti*, the counterpart of one of Walcott's original specimens of what he called *Peytoia nathorsti*. The two wider pairs of plates are at the top and bottom and at each side; between any two wider plates are seven smaller plates. Some of the teeth are visible in the lower left of the central area.

Fig. 4.76. *Anomalocaris canadensis,* an isolated, jointed limb of a specimen from Mount Stephen. Each segment has a pair of spines on the inner side. Photograph by D.E.G. Briggs.

Fig. 4.77. An isolated, jointed limb, curved and with broad, spiny blades on the inside of each segment. Limbs like this were thought by Walcott to belong to *Sidneyia* (see fig. 4.39), but are now known to belong to *Anomalocaris nathorsti* (fig. 4.79). Photograph by D.E.G. Briggs.

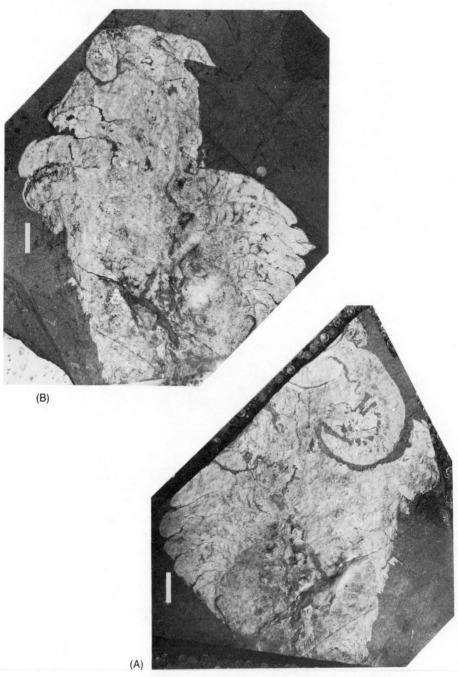

(B)

(A)

Fig. 4.78. *Anomalocaris canadensis*, the part, A, and counterpart, B, of the only almost complete specimen known, collected by the Geological Survey of Canada party from the Raymond quarry, and photographed under alcohol. The pair of appendages, the right one completely exposed, is visible in the part. The counterpart is more complete and shows the head region, which is missing from the part.

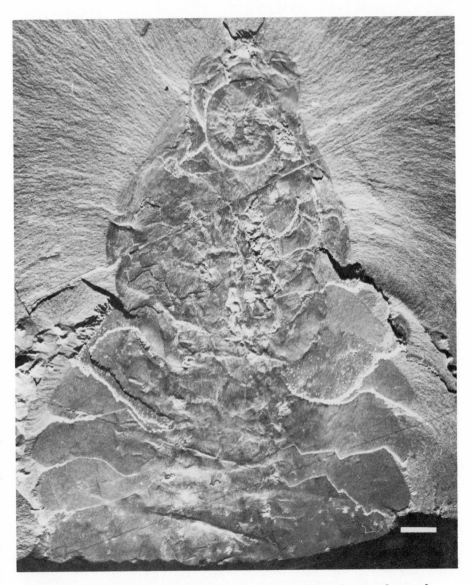

Fig. 4.79. *Anomalocaris nathorsti.* At the front are parts of a pair of appendages like those in fig. 4.77 and a circlet of plates like that in fig. 4.75. The
triangular lateral lobes overlap, and the edges of two have been exposed.
The body is incomplete posteriorly.

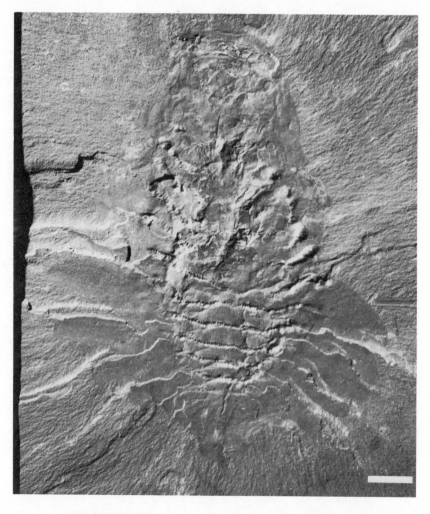

Fig. 4.80. *Anomalocaris nathorsti,* a complete specimen showing the back portion of the body, which underwent distortion as it was flattened.

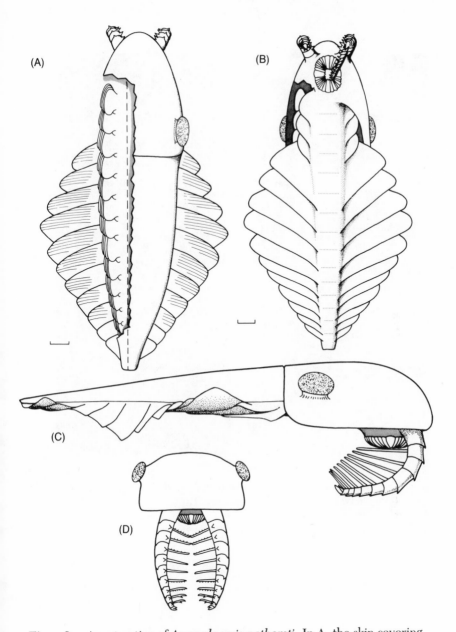

Fig. 4.81. A restoration of *Anomalocaris nathorsti*. In A, the skin covering the left side has been partly cut away to show the gills below; only a few of the many thin sheets forming the gill are shown. Stiffening rays cross the lateral lobes. The dashed line is the left margin of the alimentary canal. In B, a view from underneath, the limb on the left side of the drawing is incomplete so that the circlet of plates of the jaw may be seen. Part of the ventral covering is cut away to show the semicircular flaps and gills of the head region. In C, the view from the right side, the lobes are shown moving up and down in a wave, enabling the animal to swim over the bottom in search of prey. View D, from the front, shows the limbs in position for catching prey, and the blades and spines that project on the inner side of the limbs. Scale bar 1 cm, indicating the size of the largest complete specimens known. Restoration by H.B. Whittington and D.E.G. Briggs.

(A)   (B)

Fig. 4.82. *Amiskwia sagittiformis*, low-angle, A, and reflected, B, photographs of one of Walcott's original specimens. The tentacle (t), lateral (L) and posterior (P) fins, intestine (i), ganglion (g), and nerve chord (n) are labelled. Photographs by Conway Morris.

(A)

(B)

Fig. 4.83. *Nectocaris pteryx*, low-angle, A, and reflected, B, photographs of the only known specimen. The eye (e) is strongly reflective, above it the appendages project, and behind it is the oval shield (s). Above and below the body the short, fine parallel lines are the fin rays (r). Scale bar 0.5 cm. Photographs by Conway Morris.

5 *The Burgess Shale Fauna and Flora and Its Evolutionary Significance*

The previous chapter has shown the marvellous preservation of the animals of the Phyllopod bed, such that they can be reconstructed and modelled in detail. An early conception of what the animals may have looked like in their original site was painted by Charles R. Knight, an artist who in 1942 presented one of the earliest series of paintings in colour of life through the ages. His picture included jelly-fish (no doubt based on *Peytoia*), sponges, seaweeds, worms, and several kinds of arthropods. A diorama in the US National Museum of Natural History, Washington, DC, in the 1960s, modelled a similar scene showing the alga *Marpolia*, the sponges *Vauxia* and *Choia*, a jelly-fish, a polychaete worm, and various arthropods—trilobites, *Canadaspis*, *Marrella*, and *Sidneyia*. The new diorama (fig. 5.1), unlike the old one, has the animals modelled in life-size, and is based on a drawing

Fig. 5.1. Diorama at the US National Museum of Natural History, Smithsonian Institution, Washington, DC, showing the benthic community of the Phyllopod bed of the Burgess Shale. In the background is the vertical wall of the submarine escarpment, attached to which are growing the algae *Margaretia* (1) and *Marpolia* (2) and two species of the sponge *Vauxia* (3). Attached to the muddy bottom in the foreground is the sponge *Pirania* (4), while on or above it are the trilobite *Olenoides* (5), the sponge *Choia* (6), and, among others, the animals *Sidneyia* (7), *Canadaspis* (8), *Waptia* (9), *Opabinia* (10), and *Hyolithes* (11). Shown on the right is a hollow in the surface of the mud left by an old slump, and in the extreme foreground a vertical section through the mud revealing the worms *Ottoia* (12) and *Louisella* (13) in their burrows. Photograph reproduced by permission of the Smithsonian Institution.

that Conway Morris and I published in 1979. The diorama shows what Conway Morris has named the *Ottoia–Marrella* benthic community, which lived on, just above, or in the sea floor during deposition of the Phyllopod bed. Its members included the algae, all the sponges and brachiopods, the sea-anemone *Mackenzia* (standing upright, the basal part in the mud), the molluscs *Scenella* and *Hyolithes*, almost all the priapulid and annelid worms, trilobites except agnostoids, almost all the other arthropods, the echinoderms, supposed hemichordates and probably the chordate *Pikaia*, and the miscellaneous animals *Anomalocaris*, *Dinomischus*, *Hallucigenia*, *Opabinia*, and *Wiwaxia*. Worms that dwelt in a burrow included the polychaetes *Peronochaeta* and *Burgessochaeta*, and the priapulids *Ancalagon*, *Ottoia*, *Selkirkia*, and *Louisella*. In Walcott's collection *Ottoia* is the most numerous of these dwellers in the mud, and *Marrella* of animals that lived on or near the bottom, hence the name of the community. The records of the level within the Phyllopod bed at which each specimen was collected, made by the Geological Survey of Canada party, show that *Marrella* and *Ottoia*, as well as the worm *Selkirkia* and the trilobites *Olenoides* and *Naraoia*, occur throughout the Phyllopod bed. Many of the other arthropods are abundant in the lower half of the bed. Walcott's collection is not labelled as to exact level within the Phyllopod bed but associations of genera on single slabs of rock suggest that this community persisted throughout the time taken to deposit the bed. There are four thin layers within the bed that yielded the best and most numerous fossils, the assemblage of genera in each not exactly the same. The slump or slumps that formed a particular thin layer may have come from different sources, in each of which the composition of the fauna was slightly different. The abundant specimens of the algae *Morania* and *Marpolia* occur crowded in particular and separate layers, in which animals are found very rarely. It may be that the algae were brought in from environments in shallower and better lighted waters.

Some animals that unquestionably belonged to the benthic community, such as *Dinomischus*, are extremely rare. Other very rare animals, however, seem by their morphology to have been swimmers in higher-water layers, perhaps near the surface of the ocean. These include the polychaete worm *Insolicorypha*, the lophophorate *Odontogriphus*, the arthropods *Isoxys* and *Odaraia*, and the animals *Amiskwia* and *Nectocaris*. Agnostoid trilobites (*Ptychagnostus*, *Peronopsis*) are also regarded as pelagic (open ocean drifters and swimmers near the surface). The rarity of these pelagic animals is attributed to the infrequency with which their carcasses came to rest in the pre-slide environment. Not all are rare (less than ten specimens recovered), however; twenty-nine specimens of *Odaraia* are known, carapaces of *Isoxys* are not rare, and exoskeletons of agnostoids are quite common. Perhaps these empty shells accumulated in the pre-slide environment. Another problem is presented by the abundance of *Eldonia* (see fig. 4.63), regarded as a pelagic holothurian, in one thin layer accompanied by few other animals. Perhaps a swarm of *Eldonia* were carried in close over the escarpment, and a resulting accumulation of carcasses on a particular patch of sea floor provided the pre-slide environment. Conway Morris has drawn attention to these pelagic animals, and regards them as constituting a distinct community, which he named the *Amiskwia–Odontogriphus* community.

This discussion of communities refers to the fossils found in the Phyllopod bed in Walcott's quarry. In the Raymond quarry, 21 m higher in the Burgess Shale section (and hence a few millions of years younger), the assemblage of fossils is much more limited, and preservation less fine. No species is unique to this quarry, but whereas, for example, the sponge *Vauxia* and the arthropods *Leanchoilia* and *Sidneyia* are found, as well as the worms *Ottoia* and *Selkirkia*, *Marrella* is not. Hence the composition of the community is different at this different time and place, but there are no new soft-bodied animals. The same is true of the communities at the various localities nearby and to the south discovered by Collins, Briggs, and Conway Morris.

Figure 5.2 shows the relative proportions of numbers of genera in different groups, and the proportion of genera in each group that had mineralized hard parts. The latter reveals that the Burgess Shale contains species belonging to genera of trilobites, sponges, brachiopods, echinoderms, and molluscan shells (including hyolithids), the same five major groups of fossils with hard parts characteristic of Cambrian faunas throughout the world. What the exceptional preservation adds to this normal content is an astonishing variety of soft-bodied arthropods and other animals. In terms of numbers of genera, this is about three-fifths (64 genera) of the total (107 genera), a proportion without hard parts (and thus unlikely to be preserved as fossils) similar to that found in present-day faunas on continental shelves. A count by Conway Morris of numbers of specimens of each kind in the Walcott collection (for example, 15,000 of *Marrella*, 19 of *Aysheaia*), gave an idea of the relative numbers of individuals in the original populations. It showed that soft-bodied animals were overwhelmingly dominant in numbers, and therefore in mass of living animals. All the main types

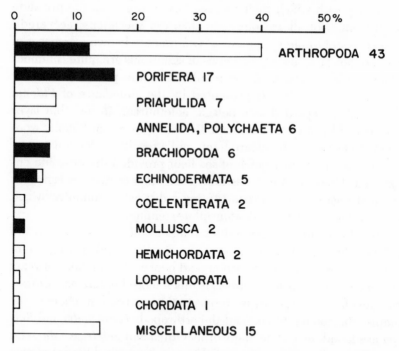

Fig. 5.2. Relative abundance of major groups of animals in the Phyllopod bed of the Burgess Shale. The length of the bar indicates the percentage of the total (107 genera) for each group, and the solid portion shows the proportion of genera in that group that had mineralized hard parts. The numbers of genera correspond with those in the list of species given in the Appendix, except that the arthropod genus 'Teles' is omitted because it may not be valid, and 'Platydendron' because it is indeterminate.

of feeding known in marine animals today may be identified—on particles in suspension in the water (brachiopods, sponges, *Dinomischus*), by ingesting mud containing organic particles (*Naraoia* has a mud-filled gut), by predation (*Olenoides, Sidneyia, Ottoia,* and *Anomalocaris*), by scavenging (*Hallucigenia* and probably some of the arthropods), and possibly by grazing (*Wiwaxia*?). The new investigation has shown that the animals of the Phyllopod bed were more varied than Walcott's published work indicated, also that they are more complex and stranger than previously believed. The importance of the discovery of this fauna, then as now unique to Cambrian rocks of the world, lies in its revelation of the nature of marine life 530 million years ago. These extraordinary fossils cannot be dismissed as some freak of evolution that occurred in a strange and isolated environment.

It is only the preservation that is unusual, not the animals, which were characteristic of the time and place. The submarine scarp below which the benthic community lived faced out towards the ocean on one side of the continent of Laurentia (see fig. 1.1), and was open to currents and migration. Many of the genera of soft-bodied animals from the Phyllopod bed have been found not only at localities 20 km to the south, but also in Middle Cambrian strata in north and west-central Utah, 1300 km further south. Patient and long-continued collecting in Utah has yielded more and more of these fossils, described in recent years by Briggs and Robison, Conway Morris and Robison, Gunther and Gunther, and Robison. In the Lower Cambrian the appendage of the *Anomalocaris canadensis* type is known from British Columbia, southern California, and Pennsylvania, a much wider geographical distribution. There is no exceptional preservation to tell us whether the Burgess fauna spread beyond Laurentia, or whether different but equally varied faunas inhabited the waters around Gondwanaland or parts of Eurasia. The soft-bodied faunas of today's seas could not be understood from so few samples coming from within a limited area around one continent, and it is unlikely that these samples show us all the kinds of animals that lived in the Cambrian seas of the world. It is much more probable that many other kinds of soft-bodied animals inhabited the early Palaeozoic seas. The rare occurrences of exceptional preservation in late Cambrian, late Silurian, or early Devonian rocks in Europe show, as might be expected, animals quite different from any in the Burgess Shale. The remarkably-preserved assemblage of minute arthropods, mostly crustaceans, recently discovered by Klaus J. Müller and his associates in Upper Cambrian rocks in Sweden, is a case in point. These fossils come from a different facies, on another continent, and have nothing in common with Burgess Shale arthropods.

As the list of species in the Appendix shows, many animals are placed in classes and phyla having living relatives, some in extinct groups such as *Hyolithes* (Hyolitha were common in the Cambrian, afterwards becoming rarer and dying out in the Permian period) and the trilobites (Cambrian to Permian). Among notable first appearances of a living group are *Mackenzia* (the earliest known soft-bodied coelenterate), the earliest barnacle *Priscansermarinus* (if it is correctly interpreted) prior to an accepted late Silurian example, the earliest crinoid and sea-cucumber, and the earliest chordate; further, '*Ottoia*' *tenuis* would be the only fossil acorn-worm (Enteropneusta) known, and *Fasciculus* the only fossil comb-jelly (Ctenophora) known, if these questionable assignments are confirmed. Malacostracan crustaceans are most important in the seas today, since they include shrimps, crabs, lobsters, and their various relatives. The phyllocarid crustaceans, so diverse in the Lower Palaeozoic, have diminished and are few in today's seas or fresh waters. Thus, as might be expected, the Burgess Shale fauna shows early members of groups that became numerous in younger rocks, as well as a variety of representatives of important Cambrian groups. Far more remarkable in the Burgess Shale, however, is the large number of miscellaneous animals that do not fit into any phylum or class of animal known today. Not only are these animals strange, but they are widely different one from another. These differences between species are typical not only of the miscellaneous animals, but also of the worms and arthropods (as the new studies have emphasized). It is rare to have more than one species of a genus, and each genus of worm or arthropod is so different from others that in most cases it seems to belong to a separate family at least. The problem this raises in classification of the arthropods has been discussed above. Why these wide differences between species and these strange 'unclassifiable' animals?

One may speculate that because the Cambrian period was the time when marine environments were first being occupied by many different animals of diverse habits, competition between species for food and space was less severe than it became subsequently. In these circumstances widely different arthropods or worms may have evolved, also strange animals that had novel combinations of structures for sensing the environment, catching or gathering food, and dealing with it. Evolution in the Cambrian was perhaps less constrained than later, and allowed for 'experiments' in living. Use of the word 'experiment' carries no necessary implication of being short-lived or rare. That an animal is preserved as a fossil is a matter of chance, even in exceptional conditions such as pertain in the Burgess Shale. Each fossil represents a population of unknown size, but some of the Burgess species were widespread geographically. Because they were preserved as fossils the miscellaneous animals must have been 'successful'—that is, there were populations of them inhabiting Cambrian seas, but in most cases we do not know how widespread geographically they were, nor how long-lived, because they are unique to the Shale. From instances of exceptional preservation in late Cambrian, Silurian, and early Devonian rocks, we know of other strange animals, including a variety of arthropods. One of these, in the lower Devonian, is an animal that appears to be related to *Marrella*, so that particular evolutionary line seems to have been long-lived. Examples of strange, unclassifiable animals are rare in the later Palaeozoic, and unknown in younger rocks. Such animals were presumably less well adapted and became extinct, while better adapted animals diversified. Thus the animals we have in today's oceans are the result not only of evolutionary expansion, but of extinctions that made way for such expansions.

Because of the variety and complexity of Cambrian life, many have argued, and still do, that there must have been a long period of multicelled animal (metazoan) evolution before the beginning of the Cambrian, perhaps as much as six hundred million years, as long as the Phanerozoic. In the last thirty years there has been intensive study of Precambrian rocks by scientists world-wide, in search of evidences of life. We now know that the earliest life, organisms of cells without nuclei (bacteria and blue-green algae or cyanobacteria), began more than three and a half thousand million years ago, about six times as long ago as the beginning of the Phanerozoic (see table 1.1). Organisms that have a cell nucleus (protists, fungi, plants, and animals) arose about one and a half million years ago, more than twice the length of time since the beginning of the Phanerozoic.

An important discovery in the Ediacara Hills of South Australia, the results of which were published in 1947, was a major stimulus to studies of the Precambrian. The fossils from the coarse, red-stained sandstones of the Ediacara Hills region are impressions, varied and complex, and have been interpreted as having been made by the soft bodies of metazoan animals that lacked hard parts. Many are circular in outline, lobate or with concentric and radial impressions, and have been considered to be jelly-fish. Also present are large (up to 1 m in length), elongate leaf-shaped forms considered to be related to modern sea-fans. Other impressions are regarded as having been made by polychaete worms; much rarer are the impressions of supposed arthropods and animals of uncertain affinities. Whether these impressions should be interpreted in terms of living coelenterates, worms, and so on, has recently been questioned, but however they are regarded, the impressions are evidence of the existence of metazoans. Fossils like those from the Ediacara Hills, in rocks older than those yielding early Cambrian shelled fossils, are now known from many localities, in Europe, North America, northern Russia, China, and South West Africa, showing that this fauna was world-wide, in rocks regarded as spanning the one hundred million years before the beginning of the Cambrian. Trace fossils, the trails made by animals moving over the surface of sediments, and burrows have been found in late Precambrian rocks and in strata as old as one thousand million years.

The earliest Cambrian rocks have also been studied intensively in recent years, in an attempt to define more precisely, at a level recognisable widely in the world, the base of the system. Limestones that are included in the Cambrian system, though below rocks containing trilobites, have yielded a remarkable variety of tiny (1–3 mm) phosphatic shells and fillings of shells, which have been extracted with acetic acid. The cap-shaped and coiled shells are generally agreed to be Mollusca, and some of the tubes may be worm tubes; but the affinities of others of these fossils are being actively debated. Trace fossils in the early Cambrian include a variety of tracks and excavations, including deep, vertical burrows.

The world-wide study of Precambrian rocks has provided ample evidence for the existence and activities of metazoan animals during the last one hundred million years of the Precambrian, and growing evidence of their activities in rocks formed up to four hundred million years before the beginning of the Phanerozoic. Seemingly there was a long period of metazoan evolution before the Cambrian, but it is only in the earliest Cambrian rocks that minute shells of metazoans appear. In a few tens of millions of years, at most, the shells of larger animals—brachiopods, trilobites, and echinoderms—are found. The Burgess Shale shows that it was not only metazoans with hard parts that were diversifying in the Cambrian, but also soft-bodied metazoans, including coelenterates, worms, arthropods, chordates, and various strange animals. The shallow seas around a number of continental masses afforded a great variety of environments for occupation, with food available from a myriad single-celled organisms. Thus, after the lapse of another hundred million years, in mid-Ordovician time, the benthic shelled marine fauna included corals, bryozoans, articulate brachiopods, cephalopods, bivalves, gastropods, crinoids, starfish and other echinoderms, and ostracodes. This fauna has a modern look, because these groups have continued to evolve, and constitute major phyla and classes of living marine benthos. Other early Palaeozoic groups, such as hyolithids and trilobites, do not survive that era, and conodonts were gone by the end of the Triassic period. The evolutionary history of the soft-bodied animals of the Burgess Shale is similar. From animals like *Canadaspis* came the varied and numerous malacostracan crustaceans, and *Aysheaia* is the kind of animal from which insects (three-quarters of all living species), centipedes, and millipedes may have arisen. The significance of *Pikaia* as the earliest known chordate and forerunner of the vertebrates, need hardly be stressed. We do not know the subsequent history of most of the strange arthropods and miscellaneous animals, since they presumably became extinct long ago. It is to be expected that such long-extinct forms will not fit into a classification of animals that is based mainly on the survivors of evolutionary diversification.

The wonderful fauna that Walcott uncovered was a tremendous and hardly believable surprise in his day, and the lessons it has for us are only gradually being understood. It does not show simply-organised animals, which might be looked on as direct ancestors of younger groups, nor animals that can be looked on as intermediates between classes or phyla. Rather the reverse, highly-organised animals in complex communities, as diverse in habit, structure, and adaptation as those of modern communities. For example, the benthic arthropods seem to have had a range of modes of feeding, including predation, scavenging, picking up or swallowing organic particles on and in the bottom mud, or filtering such particles out of the water, that was quite as varied as that seen among living crustaceans. The structure of the appendages concerned with feeding shows an equally wide range. So any notion that the Burgess Shale animals, because they come from so far in the past, were simple in either morphology, habits, or relationships within the community, is erroneous. Clues to the antecedents of this complex array, however, are few and vague, for it is very unlike the Ediacara fauna or what we know of early Cambrian animals.

The brief glimpse of marine life afforded by the Burgess Shale makes speculation about the early evolution of metazoan animals no easier. Instead it shows how little we know. Did all these animals have a common origin from a unique population of metazoan animals that first evolved in a particular place? The geography of the world before the Cambrian is difficult to decipher, but it may well be that metazoan animals arose independently in different areas. I look sceptically upon diagrams that show the branching diversity of animal life through time, and come down at the base to a single kind of animal. Metazoan animals may have originated more than once, in different places and at different times. Why did hard parts, internal and external, appear in certain (but not all) animal groups in the early Cambrian? They appear (relatively) abruptly in rocks at this time and no doubt served for protection as well as for the support of more complex muscle systems. Thus they were, in the evolutionary sense, an advantage in particular circumstances for certain kinds of animals. Did hard parts become possible because of the evolution of a particular biochemical compound, such as collagen, which is essential for linking muscles to shells? And what of the chemical and physical factors in the marine environment? When did the oceans, originally fresh water, attain the salinity they have today? How much dissolved oxygen was available? The many and complex factors bearing on animal evolution involve the physical and chemical history of the earth, so that there is immense uncertainty surrounding any speculation. Rare factual signposts such as the Burgess Shale fauna show what occurred at a particular place and time, but we are far from explaining the evolutionary pathways that coincided there, or those that lead from it.

Appendix: Species from the Phyllopod Bed, Burgess Shale[1]

ALGAE

Cyanophyta (blue-green algae)
Morania confluens Walcott, 1919
Morania elongata Walcott, 1919
Morania fragmenta Walcott, 1919
Morania? frondosa Walcott, 1919
Morania? globosa Walcott, 1919
Morania parasitica Walcott, 1919
Morania? reticulata Walcott, 1919
Marpolia spissa Walcott, 1919

Chlorophyta (green algae)
Yuknessia simplex Walcott, 1919

Rhodophyta (red algae)
Waputikia ramosa Walcott, 1919

1. This list incorporates the results of work given in the List of Publications. Because Satterthwait has published only a brief abstract of her studies on the algae, the list given here is taken from Walcott, 1919. When there is doubt about the correctness of a name, it is placed in quotation marks. Doubt regarding assignment to a particular category is indicated by a question mark following the name. The name and date following a specific name are that of the author and the date of the publication in which the species was first described; they are placed in parentheses if there has been a change in the original assignment of the species to a genus. When the species is indeterminate, the abbreviation sp. ind. follows the generic name. The list includes synonyms, that is, names originally given to specimens that have been shown subsequently to belong to a different species. Synonyms are given in parentheses, following the equals sign.

Dalyia racemata Walcott, 1919
Dalyia nitens Walcott, 1919
Wahpia mimica Walcott, 1919
Wahpia virgata Walcott, 1919
Bosworthia simulans Walcott, 1919
Bosworthia gyges Walcott, 1919

Calcareous algae
Sphaerocodium? praecursor Walcott, 1919
Sphaerocodium? cambria Walcott, 1919

Thought to be algae
Dictyophycus gracilis Ruedemann, 1931
Margaretia dorus Walcott, 1931

PHYLUM PORIFERA[2]

Class Demospongea
Vauxia gracilenta Walcott, 1920
Vauxia bellula Walcott, 1920
Vauxia densa Walcott, 1920
Vauxia venata Walcott, 1920
Leptomitus lineata (Walcott, 1920) (= *L. flexilis* (Walcott, 1920);
   L. flexilis intermedia (Walcott, 1920))
New genus for *L. bellilineata* (Walcott, 1920)
Hamptonia bowerbanki Walcott, 1920
Choia carteri Walcott, 1920
Choia ridleyi Walcott, 1920
Wapkia grandis Walcott, 1920
Halichondrites elissa Walcott, 1920
Pirania muricata Walcott, 1920
Hazelia palmata Walcott, 1920
Hazelia conferta Walcott, 1920
Hazelia delicatula Walcott, 1920
Hazelia nodulifera Walcott, 1920
Hazelia obscura Walcott, 1920
Hazelia mammillata Walcott, 1920
Hazelia? grandis Walcott, 1920
Hazelia dignata (Walcott, 1920)
Three new species of *Hazelia*
Sentinelia draco Walcott, 1920
Takakkawia lineata Walcott, 1920
Corralia undulata Walcott, 1920

---

2. This list was kindly supplied by Rigby and includes work in press.

Class Hexactinellida
Protospongia hicksi Hinde, 1888
Diagoniella hindei Walcott, 1920
New genus and species

Class Calcarea
Eiffelia globosa Walcott, 1920
New genus and species

PHYLUM BRACHIOPODA

Class Inarticulata
Lingulella waptaensis Walcott, 1924
Acrothyra gregaria Walcott, 1924
Paterina zenobia (Walcott, 1924)
Paterina pulchra (Resser, 1938)
Micromitra burgessensis Resser, 1938

Class Articulata
Nisusia burgessensis Walcott, 1924
Diraphora bellicostata Walcott, 1924

SUPER-PHYLUM LOPHOPHORATA

Odontogriphus omalus Conway Morris, 1976

PHYLUM COELENTERATA, OR CNIDARIA

Class Anthozoa
Mackenzia costalis Walcott, 1911

Fasciculus vesanus Simonetta and Delle Cave, 1978, possibly a
ctenophore

PHYLUM MOLLUSCA

Class Monoplacophora
Scenella sp. ind.

**Hyolitha—considered by many to be a separate phylum, not a class
of Mollusca**
Hyolithes carinatus Matthew, 1899

PHYLUM PRIAPULIDA

Ottoia prolifica Walcott, 1911 (= *Miskoia placida* Walcott, 1931)

Selkirkia columbia Conway Morris, 1977 (= *S. major* Walcott, 1911)

Louisella pedunculata Walcott, 1911 (= *Miskoia preciosa* Walcott, 1911)

Ancalagon minor (Walcott, 1911)

Fieldia lanceolata Walcott, 1912

Probable priapulids:

Scolecofurca rara Conway Morris, 1977

Lecythioscopa simplex (Walcott, 1931)

PHYLUM ANNELIDA

Class Polychaeta

Canadia spinosa Walcott, 1911 (= *C. irregularis* Walcott, 1911 and *C. grandis* Walcott, 1931)

Burgessochaeta setigera (Walcott, 1911)

Peronochaeta dubia (Walcott, 1911)

Insolicorypha psygma Conway Morris, 1979

Stephenoscolex argutus Conway Morris, 1979

Unidentified genus and species in Conway Morris, 1979

ARTHROPODA

Trilobita

THOSE WITH A MINERALIZED EXOSKELETON[3]

Ptychagnostus praecurrens (Westergaard, 1936) (= *Triplagnostus burgessensis* Rasetti, 1951)

Peronopsis montis (Matthew, 1899)

Pagetia bootes Walcott, 1916

Olenoides serratus (Rominger, 1887) (= *Nathorstia transitans* Walcott, 1912)

Kootenia burgessensis Resser, 1942

Oryctocephalus burgessensis Resser, 1938

Oryctocephalus matthewi Rasetti, 1951

Oryctocephalus reynoldsi Reed, 1899

Oryctocephalus sp. ind.

Parkaspis decamera Rasetti, 1951

Chancia palliseri (Walcott, 1908)

Ehmaniella waptaensis Rasetti, 1951

3. Many of these trilobites have been described by F. Rasetti; those collected by the Geological Survey of Canada are listed by Fritz.

Ehmaniella burgessensis Rasetti, 1951
Elrathia permulta (Walcott, 1918)
cf. *Elrathina brevifrons* Rasetti, 1951
Elrathina cordillerae (Rominger, 1887)
cf. *Solenopleurella* sp. ind. Rasetti, 1951
"*Solenopleurella*" sp. ind.
Hanburia gloriosa Walcott, 1916

THOSE WITH EXOSKELETON NOT MINERALIZED

Naraoia compacta Walcott, 1912 (= *N. pammon* Simonetta and
  Delle Cave, 1975; *N. halia* Simonetta and Delle Cave, 1975)
Naraoia spinifer Walcott, 1931
Tegopelte gigas Simonetta and Delle Cave, 1975

PHYLUM CRUSTACEA

Class Malacostraca, Subclass Phyllocarida
Canadaspis ovalis[4] (Walcott, 1912)
Canadaspis perfecta (Walcott, 1912) (= *Hymenocaris obliqua* Wal-
  cott, 1912; *Canadaspis obesa* Simonetta and Delle Cave, 1975)
Carnarvonia venusa[4] Walcott, 1912
Hurdia dentata[4] Simonetta and Delle Cave, 1975
Hurdia triangulata[4] Walcott, 1912
Hurdia victoria[4] Walcott, 1912
Isoxys acutangulus[4] (Walcott, 1908)
Isoxys longissimus[4] Simonetta and Delle Cave, 1975
Perspicaris dictynna (Simonetta and Delle Cave, 1975)
Perspicaris recondita Briggs, 1977
Proboscicaris agnosta[4] Rolfe, 1962
Proboscicaris ingens[4] Rolfe, 1962
Proboscicaris obtusa[4] Simonetta and Delle Cave, 1975
"*Teles*" *ovalis*[4] Simonetta and Delle Cave, 1975
Tuzoia burgessensis[4] Resser, 1929
Tuzoia canadensis[4] Resser, 1929
Tuzoia? parva[4] (Walcott, 1912)
Tuzoia praemorsa[4] Resser, 1929
Tuzoia retifera[4] Walcott, 1912

Class Ostracoda
Aluta? sp. ind.

Class Cirripedia?
Priscansermarinus barnetti Collins and Rudkin, 1981

4. Valves of carapace only are known.

Not placed in any phylum or class of Arthropoda

Actaeus armatus Simonetta, 1970
Alalcomenaeus cambricus Simonetta, 1970
Aysheaia pedunculata Walcott, 1911
Branchiocaris pretiosa (Resser, 1929)
Burgessia bella Walcott, 1912 (= *Hymenocaris? circularis* Walcott, 1912)
Emeraldella brocki Walcott, 1912 (= *Emeraldoides problematicus* Simonetta, 1964)
Habelia optata Walcott, 1912
Habelia? brevicauda Simonetta, 1964
Helmetia expansa Walcott, 1918
Houghtonites gracilis (Walcott, 1912)
Leanchoilia superlata Walcott, 1912 (= *Bidentia difficilis* Walcott, 1912; *Emeraldella micrura* Walcott, 1912; *Leanchoilia major* Walcott, 1931; *Leanchoilia amphiction* Simonetta, 1970; *Leanchoilia persephone* Simonetta, 1970; *Leanchoilia protogonia* Simonetta, 1970)
Molaria spinifera Walcott, 1912
Mollisonia symmetrica Walcott, 1912
Mollisonia rara Walcott, 1912 (= *Parahabelia rara* Simonetta, 1964)
Odaraia alata Walcott, 1912 (= *Eurysaces pielus* Simonetta and Delle Cave, 1975)
Plenocaris plena (Walcott, 1912)
Sarotrocercus oblita Whittington, 1981
Sidneyia inexpectans Walcott, 1911
Skania fragilis Walcott, 1931
Thelxiope palaeothalassia Simonetta and Delle Cave, 1975
Waptia fieldensis Walcott, 1912
Yohoia tenuis Walcott, 1912

PHYLUM ECHINODERMATA

Class Eocrinoidea

Gogia? radiata Sprinkle, 1973
eocrinoid? arm fragments

Class Crinoidea

Echmatocrinus brachiatus Sprinkle, 1973

Class Edrioasteroidea

Walcottidiscus typicalis Bassler, 1935
Walcottidiscus magister Bassler, 1936

Class Holothuroidea
Eldonia ludwigi Walcott, 1911

PHYLUM HEMICHORDATA

Class Enteropneusta?
"Ottoia" tenuis Walcott, 1911

Class Graptolithina?
Chaunograptus scandens Ruedemann, 1931

PHYLUM CHORDATA

Pikaia gracilens Walcott, 1911

MISCELLANEOUS ANIMALS[5]

Amiella ornata Walcott, 1911—a single, unique specimen
Amiskwia sagittiformis Walcott, 1911
Anomalocaris canadensis Whiteaves, 1892
Anomalocaris nathorsti (Walcott, 1911) (= *Laggania cambria* Walcott, 1911; *Peytoia nathorsti* Walcott, 1911)
Banffia constricta[6] Walcott, 1911
Dinomischus isolatus Conway Morris, 1977
Hallucigenia sparsa (Walcott, 1911)
Nectocaris pteryx Conway Morris, 1976
Oesia disjuncta[6] Walcott, 1911
Opabinia regalis Walcott, 1912
"Platydendron ovale" Simonetta and Delle Cave, 1978
Pollingeria grandis[6] Walcott, 1911
Portalia mira[7] Walcott, 1918
Redoubtia polypodia[7] Walcott, 1918
"Selkirkia" major[6] Walcott, 1908 (= *Selkirkia gracilis* Walcott, 1911)
Wiwaxia corrugata (Matthew, 1899)
Worthenella cambria[6] Walcott, 1911

---

5. The species without footnote numbers have been described in detail, with the exception of *"Platydendron ovale"*, the single specimen of which is so poorly preserved that its nature is indeterminable.

6. Originally described by Walcott as an annelid worm, but of uncertain affinities, and now under investigation by Conway Morris.

7. Originally called a holothurian by Walcott, but of uncertain affinities and now under investigation by Conway Morris.

List of Publications on the Burgess Shale

Aitken, J.D., and McIlreath, I.A. 1984. The Cathedral Reef escarpment, a Cambrian great wall with humble origins. *Geos: Energy Mines and Resources, Canada* 13(1):17–19.

Bassler, R.S. 1935. The classification of the Edrioasteroidea. *Smithsonian Miscellaneous Collections* 93:1–11.

———. 1936. New species of American Edrioasteroidea. *Smithsonian Miscellaneous Collections* 95(6):1–33.

Briggs, D.E.G. 1976. The arthropod *Branchiocaris* n. gen., Middle Cambrian, Burgess Shale, British Columbia. *Geological Survey of Canada Bulletin* 264:1–29.

———. 1977. Bivalved arthropods from the Cambrian Burgess Shale of British Columbia. *Palaeontology* 20:595–621.

———. 1978. The morphology, mode of life, and affinities of *Canadaspis perfecta* (Crustacea: Phyllocarida), Middle Cambrian, Burgess Shale, British Columbia. *Philosophical Transactions of the Royal Society, London* B 281:439–87.

———. 1979. *Anomalocaris*, the largest known Cambrian arthropod. *Palaeontology*, 22:631–64.

———. 1981. The arthropod *Odaraia alata* Walcott, Middle Cambrian, Burgess Shale, British Columbia. *Philosophical Transactions of the Royal Society, London* B 291:541–85.

———. 1981. The Burgess Shale project. US Department of the Interior, Geological Survey, open-file report 81-743, pp. 34–37.

Briggs, D.E.G., and Robison, R.A. 1984. Exceptionally preserved nontrilobite arthropods and *Anomalocaris* from the Middle Cambrian of Utah. University of Kansas Paleontological Contributions, Paper 111.

Briggs, D.E.G., and Whittington, H.B. 1981. Relationships of arthropods from the Burgess Shale and other Cambrian sequences. US Department of the Interior, Geological Survey, open-file report 81-743, pp. 38–41.

Bruton, D.L. 1981. The arthropod *Sidneyia inexpectans*, Middle Cambrian, Burgess Shale, British Columbia. *Philosophical Transactions of the Royal Society, London* B 295:619–56.

Bruton, D.L., and Whittington, H.B. 1983. *Emeraldella* and *Leanchoilia*, two arthropods from the Burgess Shale, British Columbia. *Philosophical Transactions of the Royal Society, London* B 300:553–85.

Collier, F.J. 1983. The Burgess Shale, an incredible window on the Cambrian Period. *Rocks and Minerals* 58(6):257–64.

Collins, D. 1978. A palaeontologist's paradise. *Rotunda* (Royal Ontario Museum) 11(4):12–19.

Collins, D., and Rudkin, D.M. 1981. *Priscansermarinus barnetti*, a probable lepadomorph barnacle from the Middle Cambrian Burgess Shale of British Columbia. *Journal of Paleontology* 55:1006–15.

Collins, D., Briggs, D.E.G., and Conway Morris, S. 1983. New Burgess Shale Fossil sites reveal Middle Cambrian faunal complex. *Science* 222:163–67.

Conway Morris, S. 1976a. A new Cambrian lophophorate from the Burgess Shale of British Columbia. *Palaeontology* 19:199–222.

———. 1976b. *Nectocaris pteryx*, a new organism from the Middle Cambrian Burgess Shale of British Columbia. *Neues Jahrbuch für Geologie und Paläontologie*, Monatshefte 12:705–13.

———. 1977a. Aspects of the Burgess Shale fauna, with particular reference to the non-arthropod component (abstract). *Journal of Paleontology* 51 (2 supplement):7–8.

———. 1977b. A new metazoan from the Cambrian Burgess Shale, British Columbia. *Palaeontology* 20:623–40.

———. 1977c. A new entoproct-like organism from the Burgess Shale of British Columbia. *Palaeontology* 20:833–45.

———. 1977d. A redescription of the Middle Cambrian worm *Amiskwia sagittiformis* Walcott from the Burgess Shale of British Columbia. *Paläontologische Zeitschrift* 51:271–87.

———. 1977e. *Fossil priapulid worms.* Special papers in Palaeontology, Palaeontological Association, London, 20.

———. 1978. *Laggania cambria* Walcott: a composite fossil. *Journal of Paleontology* 52:126–31.

———. 1979a. Middle Cambrian polychaetes from the Burgess Shale of British Columbia. *Philosophical Transactions of the Royal Society, London* B 285:227–74.

———. 1979b. The Burgess Shale (Middle Cambrian) fauna. *Annual Review of Ecology and Systematics* 10:327–49.

———. 1979c. Burgess Shale. In *Encyclopedia of Paleontology*, ed. R.W. Fairbridge and D. Jablonski, pp. 153–60. Stroudsburg, Pennsylvania: Dowden, Hutchinson, and Ross.

———. 1981. The Burgess Shale fauna as a mid-Cambrian community. US Department of the Interior, Geological Survey, open-file report 81-743, pp. 47–49.

———. 1982. *Wiwaxia corrugata* (Matthew). A problematical Middle Cambrian animal from the Burgess Shale of British Columbia. In *Proceedings of the Third North American Paleontological Convention, Montreal*, ed. B. Mamet and M.J. Copeland, vol. 1, pp. 93–98.

———. 1985. The Middle Cambrian metazoan *Wiwaxia corrugata* (Matthew) from the Burgess Shale and *Ogygopsis* Shale, British Columbia, Canada. *Philosophical Transactions of the Royal Society, London* B 307:507–82.

Conway Morris, S., and Robison, R.A. 1982. The enigmatic medusoid *Peytoia* and a comparison of some Cambrian biotas. *Journal of Paleontology* 56:116–22.

Conway Morris, S., and Whittington, H.B. 1979. The animals of the Burgess Shale. *Scientific American* 241:122–33.

Conway Morris, S., Whittington, H.B., Briggs, D.E.G., Hughes, C.P., and Bruton, D.L. 1982. *Atlas of the Burgess Shale*. Palaeontological Association, London.

Delle Cave, L., and Simonetta, A.M. 1975. Notes on the morphology and taxonomic position of *Aysheaia* (Onycophora?) and of *Skania* (undetermined phylum). *Monitore Zoologico Italiano*, n.s. 9:67–81.

Durham, J.W. 1974. Systematic position of *Eldonia ludwigi* Walcott. *Journal of Paleontology* 48:750–55.

Fritz, W.H. 1971. Geological setting of the Burgess Shale. In *Proceedings of the North American Paleontological Convention, Chicago, 1969*, vol. I, pp. 1155–70. Lawrence, Kansas: Allen.

Gunther, L.F., and Gunther, V.G. 1981. Some Middle Cambrian fossils of Utah. *Brigham Young University Geology Studies* 28: 1–87.

Hughes, C.P. 1975. Redescription of *Burgessia bella* from the Middle Cambrian Burgess Shale, British Columbia. *Fossils and Strata* (Oslo) 4:415–36.

Hutchinson, G.E. 1931. Restudy of some Burgess Shale fossils. *Proceedings of the United States National Museum* 78(11):1–24.

———. 1969. *Aysheaia* and the general morphology of the Onychophora. *American Journal of Science* 267:1062–66.

Knight, C.R. 1942. Parade of life through the ages. *National Geographic Magazine* 81:141–84.

McIlreath, I.A. 1977. Accumulation of a Middle Cambrian, deep-water limestone debris apron adjacent to a vertical, submarine carbonate escarpment, southern Rocky Mountains, Canada. Society of Economic Paleontologists and Mineralogists, Special Publication 25:113–24.

Piper, D.J.W. 1972. Sediments of the Middle Cambrian Burgess Shale, Canada. *Lethaia* 5:169–75.

Rasetti, F. 1951. Middle Cambrian stratigraphy and faunas of the Canadian Rocky Mountains. *Smithsonian Miscellaneous Collections* 116(5):1–277.

Raymond, P.E. 1920. The appendages, anatomy and relationships of trilobites. *Memoirs of the Connecticut Academy of Arts and Sciences* 7:1–169.

———. 1931. Notes on invertebrate fossils, with descriptions of new species. *Bulletin of the Museum of Comparative Zoology, Harvard University* 55(6):165–213.

———. 1935. *Leanchoilia* and other Mid-Cambrian Arthropoda. *Bulletin of* the Museum of Comparative Zoology, Harvard University 76(6):205–30.

Resser, C.E. 1929. New Lower and Middle Cambrian Crustacea. *Proceedings of the US National Museum* 76:1–18.

———. 1938. Fourth contribution to nomenclature of Cambrian fossils. *Smithsonian Miscellaneous Collections* 97(10):1–43.

Robison, R.A. 1976. Middle Cambrian trilobite biostratigraphy of the Great Basin. *Brigham Young University Geology Studies* 23(2):93–109.

———. 1982. Some Middle Cambrian agnostoid trilobites from western North America. *Journal of Paleontology* 56:132–60.

———. 1984. *New occurrences of the unusual trilobite* Naraoia *from the Cambrian of Idaho and Utah.* University of Kansas Paleontological Contributions 112, pp. 1–8.

———. 1985. Affinities of *Aysheaia* (Onychophora) with description of a new Cambrian species. *Journal of Paleontology* 59:226–35.

Rolfe, W.D.I. 1962. Two new arthropod carapaces from the Burgess Shale (Middle Cambrian) of Canada. *Breviora* (Museum of Comparative Zoology, Harvard University) 160:1–9.

Ruedemann, R. 1931. Some new Middle Cambrian fossils from British Columbia. *Proceedings of the US National Museum* 79:1–18.

Satterthwait, D.F. 1977. Paleoenvironmental implications of the Burgess Shale flora (abstract, North American Paleontological Convention II). *Journal of Paleontology* 51 (Supplement to no. 2):24.

Schuchert, C. 1927. Charles Doolittle Walcott, paleontologist, 1850–1927. *Science* 65: 455–58.

Simonetta, A.M. 1962. Note sugli artropodi non trilobiti della burgess shale, cambriano medio della Columbia Britannica (Canada). *Monitore Zoologico Italiano* 69:172–85.

———. 1963. Osservazioni sugli artropodi non trilobiti della 'Burgess Shale' (Cambriano medio). *Monitore Zoologico Italiano* 70–71: 99–108.

———. 1964. Osservazioni sugli artropodi non trilobiti della 'Burgess Shale' (Cambriano medio). III contributo. *Monitore Zoologico Italiano* 72:215–31.

———. 1970. Studies on non-trilobite arthropods of the Burgess Shale (Middle Cambrian). *Palaeontographica Italica* 66 (n.s. 36):35–45.

———. 1976. Remarks on the origin of the Anthropoda. *Atti della Società Toscana di Scienze Naturali, Memorie (1975)*, B 82:112–34.

Simonetta, A.M., and Delle Cave, L. 1975. The Cambrian non-trilobite arthropods from the Burgess Shale of British Columbia. A study of their comparative morphology, taxinomy [sic.] and evolutionary significance. *Palaeontographica Italica* 69 (n.s. 39), 1–37.

———. 1978a. Notes on new and strange Burgess Shale fossils (Middle Cambrian of British Columbia). *Atti della Società Toscana di Scienze Naturali, Memorie*, A 85:45–49.

———. 1978b. Una possibile interpretazioni filogenetica degli artropodi paleozoici. *Bollettino di Zoologia* 45:87–90.

———. 1980. The phylogeny of the Palaeozoic arthropods. *Bollettino di Zoologia* 47 (supplement):1–19.

———. 1981. An essay in the comparative and evolutionary morphology of Palaeozoic arthropods. *Atti dei Convegni Lincei, Roma* 49:389–439.

———. 1982. New fossil animals from the Middle Cambrian. *Bollettino di Zoologia* 49:107–14.

Sprinkle, J. 1973. *Morphology and evolution of blastozoan echinoderms.* Museum of Comparative Zoology, Harvard University, special publication.

Størmer, L. 1939. Studies on trilobite morphology. Part 1. The thoracic appendages and their phylogenetic significance. *Norsk geologisk tidsskrift* 19:143–273.

———. 1944. On the relationships and phylogeny of fossil and recent Arachnomorpha. *Skrifter utgitt av Det Norske Videnskaps-Akademni Oslo, I Matematisk-maturvidenskapelig klasse* 5:1–158.

Walcott, C.D. 1908. Mount Stephen rocks and fossils. *Canadian Alpine Journal* 1(2):232–48.

———. 1911a. Middle Cambrian Merostomata. Cambrian Geology and Paleontology II. *Smithsonian Miscellaneous Collections* 57:17–40.

———. 1911b. A Geologist's paradise. *National Geographic Magazine* 22:509–21.

———. 1911c. Middle Cambrian Holothurians and Medusae. Cambrian Geology and Paleontology II. *Smithsonian Miscellaneous Collections* 57:41–68.

———. 1911d. Middle Cambrian Annelids. Cambrian Geology and Paleontology II. *Smithsonian Miscellaneous Collections* 57:109–44.

———. 1912a. Middle Cambrian Branchiopoda, Malacostraca, Trilobita and Merostomata. Cambrian Geology and Paleontology II. *Smithsonian Miscellaneous Collections* 57:145–228.

———. 1912b. Cambrian of the Kicking Horse Valley, British Columbia. Summary Report Geological Survey Branch, Department of Mines, Canada, 1911, Sessional Paper, 26:188–91.

———. 1912c. Studies in Cambrian geology and paleontology in the Canadian Rockies. *In* Expeditions organized or participated in by the Smithsonian Institution in 1910 and 1911. *Smithsonian Miscellaneous Collections* 59:39–45.

———. 1912d. Cambrian Brachiopoda. US Geological Survey Monograph 51: pt. 1, 1–872; pt. 2, 1–363.

———. 1913. Geological exploration in the Canadian Rockies. *In* Explorations and fieldwork of the Smithsonian Institution in 1912. *Smithsonian Miscellaneous Collections* 60:24–31.

———. 1914. Geological explorations in the Canadian Rockies. *In* Explorations and fieldwork of the Smithsonian Institution in 1913. *Smithsonian Miscellaneous Collections* 63:2–12.

———. 1917. Story of Granny, the mountain squirrel. *Journal of Animal Behaviour* 7:454–55.

———. 1918a. Geological explorations in the Canadian Rockies. *In* Explorations and fieldwork of the Smithsonian Institution in 1917. *Smithsonian Miscellaneous Collections* 68:4–20.

———. 1918b. Appendages of trilobites. Cambrian Geology and Paleontology IV. *Smithsonian Miscellaneous Collections* 67:115–216.

———. 1919. Middle Cambrian Algae. Cambrian Geology and Paleontology IV. *Smithsonian Miscellaneous Collections* 67:217–60.

———. 1920. Middle Cambrian Spongiae. Cambrian Geology and Paleontology IV. *Smithsonian Miscellaneous Collections* 67:261–364.

———. 1921. Notes on structure of *Neolenus*. Cambrian Geology and Paleontology IV. *Smithsonian Miscellaneous Collections* 67:365–456.

———. 1924. Cambrian and Ozarkian Brachiopoda. Cambrian Geology and Paleontology IV. *Smithsonian Miscellaneous Collections* 67:477–554.

———. 1931. Addenda to descriptions of Burgess Shale fossils. *Smithsonian Miscellaneous Collections* 85:1–46 (with explanatory notes by Charles E. Resser).

Walcott, S.S. 1971. How I found my own fossil. *Smithsonian* 1(12):28–29.

Walton, J. 1923. On the structure of a Middle Cambrian alga from British Columbia (*Marpolia spissa* Walcott). *Proceedings of the Cambridge Philosophical Society, Biological Sciences* 1:59–62.

Whittington, H.B. 1971a. The Burgess Shale: History of research and preservation of fossils. In *Proceedings of the North American Paleontological Convention, Chicago, 1969*, part I, pp. 1170–1201. Lawrence, Kansas: Allen.

———. 1971b. Redescription of *Marrella splendens* (Trilobitoidea) from the Burgess Shale, Middle Cambrian, British Columbia. *Geological Survey of Canada Bulletin* 209, 1–24.

———. 1974. *Yohoia* Walcott and *Plenocaris* n. gen., arthropods from the Burgess Shale, Middle Cambrian, British Columbia. *Geological Survey of Canada Bulletin* 231, 1–21 (figs 1–6 of plate X should be interchanged with figs 1–5 of plate XII).

———. 1975a. The enigmatic animal *Opabinia regalis*, Middle Cambrian, Burgess Shale, British Columbia. *Philosophical Transactions of the Royal Society, London*, B 271:1–43.

———. 1975b. Trilobites with appendages from the Middle Cambrian, Burgess Shale, British Columbia. *Fossils and Strata* (Oslo) 4:97–136.

———. 1977. The Middle Cambrian trilobite *Naraoia*, Burgess Shale, British Columbia. *Philosophical Transactions of the Royal Society, London*, B 280:409–43.

———. 1978. The lobopod animal *Aysheaia pedunculata* Walcott, Middle Cambrian, Burgess Shale, British Columbia. *Philosophical Transactions of the Royal Society, London* B 284:165–97.

———. 1979. Early arthropods, their appendages and relationships. In *The Origin of Major Invertebrate Groups*, Systematics Association, Special Volume 12, ed. M.R. House, pp. 253–68. London: Academic Press.

———. 1980a. Exoskeleton, moult stage, appendage morphology and habits of the Middle Cambrian trilobite *Olenoides serratus*. *Palaeontology* 23:171–204.

———. 1980b. The significance of the fauna of the Burgess Shale, Middle Cambrian, British Columbia. *Proceedings of the Geologists' Association* 91:127–48.

————. 1981a. Cambrian animals: Their Ancestors and Descendants. *Proceedings of the Linnean Society, New South Wales* 105:79–87.

————. 1981b. Rare arthropods from the Burgess Shale, Middle Cambrian, British Columbia. *Philosophical Transactions of the Royal Society, London* B 292:329–57.

————. 1982. The Burgess Shale fauna and the early evolution of metazoan animals. In *Palaeontology, Essential of Historical Geology*, ed. E.M. Gallitelli, STEM Mucchi, pp. 11–24. Modena.

————. 1985. *Tegopelte gigas*, a second soft-bodied trilobite from the Burgess Shale, Middle Cambrian, British Columbia. *Journal of Paleontology*.

Whittington, H.B., and Briggs, D.E.G. 1982. A new conundrum from the Middle Cambrian Burgess Shale. In *Proceedings of the Third North American Paleontological Convention, Montreal*, ed. B. Mamet and M.J. Copeland, vol. 2, pp. 573–75.

————. In press. The largest Cambrian animal, *Anomalocaris*, Burgess Shale, British Columbia. *Philosophical Transactions of the Royal Society, London*, B 309:569–609.

Yochelson, E.L. 1961. The operculum and mode of life of *Hyolithes*. *Journal of Paleontology* 35:152–61.

————. 1967. Charles Doolittle Walcott, 1850–1927. *Biographical Memoirs, National Academy of Sciences of the United States* 39:471–540.

Index

149

'Ottoia,' 69, 127

part (of specimen), 47, 60, 66
pelagic animals, 124–25
Peronochaeta, 124
Peronopsis, 124
Perspicaris, 67
Peytoia, 53, 72–74, 122
Phyllocarida, 43, 60–62, 67, 128
Phyllopod bed, 18–19, 27, 35, 36, 37,
     38, 39, 41, 43, 122, 124, 125, 126, 127
phylum, 42, 45
Pikaia, 54, 70, 124, 130
Piper, David J.W., 33
Pirania, 50, 51
Plenocaris, 61
Precambrian fossils, 53, 129–30
predator, predation, 54, 58, 62, 66, 74,
     75, 126, 131
Priscansermarinus, 127
Proboscicaris, 43
Protospongia, 51, 66
Ptychagnostus, 44, 124
pyrite, 35, 69

Raymond, Percy E., xiv, 17, 22, 46, 64,
     71, 72
Raymond quarry, 17, 18, 20, 22, 41, 125
Reed, F.R.C., 43
reef, 7, 8, 9, 26–27
Resser, C.E., 16, 46
Reynolds, S.H., 43
Rigby, J. Keith, xv, 49, 50
Robison, Richard A., 44, 127
Royal Ontario Museum party, 22, 71
Ruedemann, Rudolf, 46
Rust, William P., 13

Sarotrocercus, 63, 68
Satterthwait, Donna L., 49, 133 n
scavenger, scavenging, 5, 35, 56, 57, 58,
     72, 126, 131
Scenella, 36, 54, 124
Schuchert, Charles, 11
scolecodonts, 56
sea-anemone, 53, 124
sea-cucumber (Holothuroidea), 9, 53,
     69, 125, 127
sea urchins, 8, 68
Selkirkia, 36, 55, 124, 125
Sidneyia, 54, 62–63, 68, 73, 122, 125,
     126
silicate of alumina and calcium, 36

Simonetta, Alberto M., 46, 53, 64
slump, submarine, 30–33, 38, 50, 60,
     124
snails (gastropods), 5, 6, 9, 72, 130
species: definition, 42–43; author and
     date, 43
spiders, 67
split: revealing fossil, 38, 47
sponges, 6–7, 16, 33, 49, 66, 69, 122,
     124, 125, 126
Sprinkle, J., xv, 69
starfish, 8, 130
Stephen Formation, 22, 26–27, 38, 40
Størmer, Leif, 46, 68
stromatolites, 8

Takakkawia, 50
Tegopelte, 59, 67
trace fossils, xiii, 35, 58, 129–30
trilobites, 5, 9, 26, 36, 43–44, 62, 67, 74,
     124, 125, 127, 130; soft-bodied, 59
Triplagnostus, 44
turbidite deposit, 33, 35
Tuzoia, 43

U.S. National Museum of Natural His-
     tory, xiii, 12, 15, 17; diorama, 22, 122–
     24
Utah, 127

Vauxia, 34 (fig. 3.8), 45, 50, 122, 125

Walcott, C.D., camp site, 21, 22
Walcott, Mrs. Mary Vaux, 16, 21, 45
Walcott, Sidney S., 62
Walcott collection, xiii, 12, 17, 42, 46,
     60, 124, 125
Walcott quarry, 12, 17, 18–23, 33, 38,
     39, 40, 41, 43, 64, 68, 72, 125
Walcottidiscus, 68
Walton, J., 49
Wapta Mountain, 11, 18, 23, 40, 45
Waptia, 45, 65, 68
Wiwaxia, 49, 54, 70, 72, 124, 126
worms, 9, 16, 33, 36, 38, 44, 49, 57, 70,
     71, 72, 74, 122, 124, 128, 129, 130;
     acorn, 127; annelid, 54, 124; arrow, 74;
     polychaete, 38, 55–56, 66, 122, 124,
     129; priapulid, 55–56, 124; ribbon, 74

Yoho National Park, 22
Yohoia, 34 (fig. 3.8), 65